COSMIC
CATASTROPHES

GERRIT L. VERSCHUUR

ADDISON-WESLEY PUBLISHING COMPANY

Reading, Massachusetts   Menlo Park, California
London   Amsterdam   Don Mills, Ontario   Sydney

Library of Congress Cataloging in Publication Data

Verschuur, Gerrit L., 1937-
    Cosmic catastrophes.

    1. Astronomy.   2. Natural disasters.   3. Earth
sciences.   I. Title.
QB52.V47   520   77-92164
ISBN 0-201-08098-2
ISBN 0-201-08099-0 pbk.

Illustrations on pages 4, 39, 51, 61, 70, 77, 111, 134, 147, 160, 179, 188, 203 © 1978 by Stephen Rinn Fundingsland.

ISBN 0-201-08098-2-H
ISBN 0-201-08099-0-P
ABCDEFGHIJK-AL-798

Dedicated
to the memory of
Steve Fundingsland,
our friend

ACKNOWLEDGEMENTS

During the last several years an increasing number of articles in the astronomical and related literature have dealt with aspects of the universe that have a bearing on our existence on earth. Without the solid foundations of the basic research that went into those articles, a book such as this could not have been created. I am therefore indebted to those many astronomers and other scientists whose work is drawn upon in so many ways.

An early version of this book was read by several people and I am particularly grateful to Susan Lynds, Don Goldsmith and Steve Fundingsland for helpful comments. "I also want to thank Geri Atkins for her editorial contributions, and the staff of Addison-Wesley. Finally, I wish to express my gratitude to Steve Fundingsland for permission to use his artwork and making available the half-tone photographs of these original paintings for use as illustrations in the text.

CONTENTS

Introduction vii

1 Death Star* 1

2 Starbirth to Stardeath 16

3 The Good Hope* 32

4 Supernovae and Life 50

5 Magnetic Field Zero 65

6 Ice Age* 74

7 Our Sun, Our Galaxy, and Ice Ages 91

8 Impact* 108

9 Craters Everywhere 118

10 The Black Hole Incident* 130

11 Black Holes 143

12 Red Giant* 155

13 How to Cope with Catastrophe 177

14 Contact* 186

15 Our Future on Earth 199

Index 209

*Science fiction

INTRODUCTION

This is a book that deals with catastrophe. And does so entertainingly.

More than that, Professor Gerrit Verschuur uses the techniques of science fiction to describe catastrophic events in human terms. How would *you* react to the inexorable freezing of our entire planet if we should enter a new ice age? What would *you* do if you suddenly learned that the explosion of a star some fifty trillion miles away was going to force you and the next ten generations of your grandchildren to live underground?

Impossible? No. Unlikely? Certainly—but so were Hiroshima and the *Titanic's* sinking. Do you honestly believe that nature cannot produce cataclysms that dwarf the work of mere human beings?

To me, the thing that makes this book so vivid is Professor Verschuur's use of fiction as well as nonfiction to make these cosmic catastrophes come alive for the reader. It's one thing to read that our sun may evolve into a red giant someday. It's quite another to witness human beings just like us trying to cope with the world-shattering implications of that cosmic event.

Using the techniques of fiction to give human dimensions to cosmic events is an old standby of science fiction writers. In this book, Verschuur uses science fiction to show the catastrophes that may lurk in our future. Then he gives the scientific facts that make each scenario a plausible and chilling possibility.

Science fiction is very much in vogue these days. Universities and high schools offer courses in it. Major motion picture studios invest millions of dollars in films about flying saucers and interstellar wars. Nearly ten percent of all the fiction books published in the United States are science fiction. Yet very few people know how to use this tricky medium to get across a vital message.

Many people believe that science fiction predicts the future. After all, there have been science fiction stories predicting submarines, spacecraft, nuclear weapons, lasers, artificial organs, undersea habitats, and countless other technological marvels. But these stories have also predicted invisibility, interstellar travel, immortality, visits from alien beings, travel through time, weather control, mental telepathy, and many other wonders that seem unlikely if not downright impossible.

Science fiction predictions, then, are something like a broken clock. After all, a broken clock is correct twice each day. There have been so many science fiction stories full of "predictions" that some were bound to come true.

Moreover, while science fiction writers correctly predicted that automobiles would lead to urban sprawl, they were much too late in examining the consequences of traffic snarls and air pollution. Medical wonders dot the pages of science fiction stories, yet which science fiction writer predicted a worldwide population explosion as the result of modern medicines? And though literally hundreds of science fiction stories predicted the first manned landing on the moon, not one of them foresaw that this bit of history would be televised live to billions of viewers back on earth.

Then what is science fiction, if it is not predictive?

As Verschuur shows in this book, science fiction is a finely honed tool for presenting *alternatives,* glimpses of possible futures, examinations of what may happen tomorrow if certain assumptions about the future are granted.

In this sense, science fiction serves as a sort of social test bench, the kind of "analog model" that computer scientists set up to check out theories in physics or chemistry. If you start with situation A and add ingredient B, what happens? If you start with the world the way it is today and explode a supernova some 30 lightyears away, what happens to the human race?

Science fiction writers, then, are not attempting to predict the future. Most of them do not believe that there is such a thing as The Future, immutable and foreordained. The future lies open before us, a tantalizing myriad of possibilities. Science fiction can show us where some of these possibilities might lead and can serve to guide our future actions.

We do not live in a mechanistic universe that marches blindly into a preordained future. If quantum physics and the Uncertainty Principle have taught us anything, it is that the future—even a picosecond ahead—is open to a nearly infinite variety of possibilities.

Seen in this light, it becomes apparent that science fiction writers—Verschuur among them—are a rather strange breed of prophet. They are not saying, "This will happen." Rather, they warn, "This *could* happen. What will you do if it does?"

Scant wonder many people loathe science fiction! It forces them to think, to face the reality that change is inevitable, that tomorrow will not only be different but perhaps a good deal scarier than today.

By a curious twist of semantics, science fiction has been branded "escape literature." Escape? Into reality! Every other form of literature is escapist—it deals *exclusively* with the past. You can read every word ever written about the Civil War or Vietnam and you will never be able to change a moment of those events. But science fiction puts a sort of special responsibility on its readers: since it describes potential futures, it tacitly forces us to consider what we should *do* about these potential tomorrows.

In days of old, prophets often ran the risk of being treated roughly when their prophecies were unpleasant. There is an old tradition of blaming the messenger for the bad news.

Professor Verschuur shows us some very bad news indeed: a half-dozen ways in which our lives, our civilization, our entire species can be snuffed out forever. One is tempted to laugh, or at least smile, at such prophecies. Perhaps by disbelieving them we can prevent them from coming true. But, although none of these doomsdays may ever come to pass, the statistical chances are that *at least one of them will indeed happen.*

What will we do then?

Science fiction has another ennobling quality: it teaches us to take the long view of things.

Other human institutions are essentially backward-looking: the law, religion, social customs, government—all are dedicated to maintaining the status quo, to keeping everything just as it was yesterday. Science, by its very nature, is constantly uncovering new ideas, new forces, new opportunities, new dangers. It is thoroughly forward-looking, and it demands change.

Science fiction is the expressive, often serious voice of science. In a society where "future planning" means looking as far ahead as the next weekend, the next vacation, or (for far-seers) the next political elections, science fiction forces us to think in terms of centuries, even millennia. We read it and ponder not merely individual people or nations but our whole planet and our species.

If you read this book carefully, you will see that there is a unity to the universe. We are the spawn of stars, and into stardust we shall return.

Cosmic catastrophes have struck our world before. We have weathered supernova explosions, ice ages, magnetic field reversals, and other cataclysms.

But if we are to weather the disasters that lie in our future, we must take the long view, the unbiased view, and face these enormous problems as a united species. For, as this book tells us, these cosmic catastrophes will overtake us all—regardless of nationality, race, or economic status. The entire human race will be threatened. Only the concerted, coordinated actions of the entire human race can lead to our continued survival.

Ben Bova

Manhattan
January 1978

DEATHSTAR
CHAPTER ONE

Committee on Alien History

Project Sol, Report, #17

Effect of Supernova SN-24-01 on the Planet Earth

Based on research into the Terrestrial Microfilm Archives, including diaries, newspapers, and
government documents.

Monat 10 (2537 Earth Time)

It appears the supernova was first seen in 1987 earth
time, precise date uncertain. People outside on that
summer evening slowly became aware of a bright light
among the stars. Newspaper offices and observatories as
well as police stations and airports were immediately
flooded with phone calls from the curious and alarmed.
   Amateur and professional astronomers studied the
sky, but that didn't help much because the light, which
grew brighter by the hour, was blinding when viewed
through a telescope. Soon the night sky was lit as if by
a full moon, and thousands jammed the streets to

watch. In some places panic started. Wild rumors circulated about an attack from outer space, but astronomers interviewed on radio and TV stated that the light almost certainly came from a supernova. Before they could be sure, and predict what would happen next, they had to find out what star had exploded. Daniel Ferguson, an astronomer in Chicago, was the first of many to measure the location of the object in the sky and then check that position in a star catalog.

He told the news media that it was a star well known and charted. The interesting thing, he said, was that current knowledge of supernovae and the brightness of this star before the explosion indicated that it could easily end up producing as much light as the sun itself. Many astronomers reached the same conclusion, and word spread that the star would outshine the sun in a day or two.

Few people went to bed that night as they watched the new star grow brighter. Every available telescope was used to study the spectrum and radio intensity of the star. X-ray telescopes on satellites that were directed toward the supernova promptly burned out their detectors.

On the streets, in homes, bars, universities, shopping centers, everyone was asking, "What will happen next?" Suddenly all astronomers, professional or amateur, were experts on the subject.

When the first data were analyzed at several of the major national observatories, it was quite apparent from the spectrum of the star that it was indeed a supernova. For many years astronomers had expected one to occur somewhere in the galaxy, but this one was too close for comfort. Only 30 light years away. Earlier studies of the effects of supernovae were quickly dug up from scientific journals, and in the early hours of the morning the astronomers checked their calculations. The supernova would probably produce a serious depletion of the ozone layer. But when? And how extensive would the damage be?

Checks with cosmic ray observatories revealed that no increase in cosmic rays had been detected. This was hardly surprising, since cosmic rays would have to wend an erratic path through the magnetic fields between stars in order to reach earth. How long it would take to do this depended on the structure and strength of the fields between the supernova and earth, and no one knew what either of those were.

How much should be revealed to the public? Word was sent to the President that a select group of astronomers would like to make him aware of their considerations. Ferguson and eight other scientists arrived in Washington just after midnight. In their conference with the President, they stressed that a cosmic ray blast would definitely hit sooner or later, anywhere from a year to a hundred years from 1987. The estimates for this time were as varied as the scientists performing the calculations.

Word leaked out, and the media managed to distort the facts on the way to the public so that wild stories ranging from "no harm expected" to "massive extinction of life by noon," spread around the country.

Finally, at 3:00 A.M. the President spoke. By then the night sky was as bright as an overcast day, and no one could look straight at the supernova anymore without risking eye damage. An old video tape of his speech has been found (TMA Document 28-13).

"My fellow Americans. We are witnessing a most profound demonstration of the powers of nature. [Unintelligible] Astronomers have informed me that this bright object in the sky is a supernova. They are extremely rare events. I am told that the last several supernovae, all much more distant than this one, were

Remnant Bay. Oil on canvas. The close interdependence between supernovae (distant exploding stars) and life on earth, as well as the formation of the earth itself, is illustrated here. All the material of which our planet is made, as well as the life on it, consists of atoms produced in distant supernovae a long time ago, elsewhere in the Milky Way galaxy.

seen back in 1604, 1572, and 1054. They were no brighter than some of the planets in our skies, hence no danger to us. The problem we must now confront is whether this supernova, being relatively close to the sun, will directly influence life on earth. We are considering whether to take special precautions to safeguard life in the future. [Unintelligible]

I am assured that no immediate harm will come to the people of this planet as a result of this explosion. Nevertheless, I have established a select committee to investigate what actions should be immediately taken. In the meantime, it is predicted that this new star will continue to increase in brightness for another day or two, albeit more slowly, and will then start a long, slow decline over several years until it is once again invisible except at night. The star will remain above the horizon year around here in the northern hemisphere, which means that it will always be visible at night. Our daytime sky will appear to have two suns.

There is, I repeat, no cause for immediate alarm. The major effect on our lives will be that no artificial lighting will be needed at night. Obviously, this will enormously help in the energy crisis we are now facing.

In conclusion, I have been told that a sight such as we are now experiencing is likely to be seen on earth only once in several hundred thousand years. The last time it happened is not known to us, but life on earth survived then, and we will most certainly survive now. Good morning and God bless you all!"

It is not immediately evident what he meant by this final phrase.

The committee organized under Daniel Ferguson met for the first time later that day, an unusually bright day. In what was to be the last fruitful and ordered meeting for a year, the committee delegated each of its members to set up subcommittees all over the country, to get as many points of view as possible about the aftereffects of the supernova. At a news conference later in the day, Ferguson fielded questions. He promised that the committee would monitor cosmic ray levels and would conduct public hearings before taking any action.

As a simple precaution the military immediately set about checking all the fallout shelters in the country and resupplying them with food. This move was made secretly, with maximum security, to avoid public panic. Ferguson concurred in this covert activity. According to his diary (TMA Document 16-42), he felt the measure was justified by the need for national security.

The subcommittees were set up and quickly swamped with constructive advice, as well as opinions from doomsayers, who were given a hearing for once. Many claimed that the earth would end soon, and deep fear spread across the country.

The unavoidable presence of what was now being called the Death Star influenced everyone's life. Only in the southern hemisphere of the planet, where the new star wasn't visible, did life continue normally, or as normally as possible when the airwaves were constantly filled with tales of alarm.

The subcommittees ended their first week's work with as many different points of view as there were members. They agreed on only one underlying theme, namely, that the cosmic rays from this Death Star were going to arrive on earth sooner or later!

The Ferguson committee became the Committee for Surveillance of the Supernova, CSSN. Its activities were classified by the government as soon as the President and the military learned that the committee was airing fears for the future. Ferguson was beginning to doubt the need for secrecy, but he was overruled by the Pentagon (a military organization). Despite many statements from prominent lay persons that the unknown cosmic ray blast would arrive sooner or later, the government thought that as long as CSSN was kept quiet, the alarming stories would die down.

In some quarters, classification of CSSN's work was interpreted to mean something much worse. Wild stories abounded. According to a fanatic sect that sprang up, the new star was a gigantic alien spaceship that was slowing down as it approached the earth. Others thought the new star was moving closer to the sun and would soon collide. Large circulation newspapers; after a few days of semi-responsible reporting, went back to headlines like "The Truth Behind the New Star—The story astronomers dare not tell!"

Weeks went by as CSSN struggled to bring some order into the official chaos. Finally, it drew up a plan of action for each of several eventualities.

Eventuality A, the worst conceivable, would be the sudden onset of a cosmic ray burst. In that case, the majority of the people in countries like the United States could get into fallout shelters, but they couldn't stay there very long. According to the CSSN scientists, a Sudden Onset (SO) would probably mean that the cosmic ray dose would remain very high for a century or more. By that time, the earth's ozone layer would be completely gone, and humans wouldn't be able to venture outdoors at all. This raised the immediate problem of how a limited number of survivors could survive in shelters for 100 years or more.

Part of Plan A was a crash program to produce a 500-year survival capability for a few thousand people, including a way to produce food. Since all living things would suffer from the disaster, natural food supplies on the surface of the planet would be killed off. One solution was to construct large shelters deep inside the Rocky Mountains, where food could be grown in special greenhouses. Another possibility was escape to the bottom of the sea. This theory appeared ideal, since the sea harbored tremendous supplies of food and also provided protection from space radiation. So another crash program was implemented to study how large colonies might survive under the sea for 500 years.

The climate of the planet would, of course, be seriously altered by the removal of the ozone layer. If the polar caps melted, the sea level would rise so much that all low-lying lands would be inundated. By the time that happened, however, food supplies would have run out for most of earth's population. There was no conceivable way to avert this threat in the short term. Plan A, to provide for only a few thousand survivors, was assigned to NASA and the Navy. (Historian's note: NASA apparently refers to the Space Controllers on earth. The Navy used ships that sailed on the oceans.)

Eventuality B, according to CSSN, was that the cosmic ray blast would reach earth in about 50 years. The best way to meet this possibility was to have NASA obtain an early warning of the cosmic ray onset by launching deep space probes equipped with cosmic ray detection equipment. These space probes would have such powerful transmitters that they could be tracked for 100 years as they moved out into space. Twenty probes were to be launched, and it was conservatively estimated that they would give a few days' warning if the blast didn't reach earth within the next 50 years. The first half dozen space probes, nicknamed Armageddon A through F, were launched a year later.

Eventuality C accepted the broad belief that the cosmic ray increase would be gentle and would reach the earth in about 110 years. In that case, the depletion of the ozone layer would also be gradual, and contingency plans were made for various rates of change. The idea that eventually proved most popular was the construction of enclosed mall cities. These were regarded as

safe and adequate precautions against gradual cosmic
ray increase. Given enough time and zero population
growth, the entire population of the United States could
be housed in these enormous shelters. CSSN also
authorized studies of possible ways to replenish the
ozone layer artificially.

A year passed, and conditions didn't change much.
Darkness was still absent in the northern hemisphere, a
phenomenon that led to a widespread change in the
way of life of several billion humans. Work shifts, for
example, were spread over the 24-hour day. Even
though the sunlit day was still the favorite time for rest-
ing on weekends, the Death Star's "day" was a novel
experience that was turned to widespread commercial
and recreational use.

The public had very nearly forgotten about CSSN.
But the committee, still working hard, was coming to
the unshakable conclusion that the cosmic rays would
come sometime between 10 and 100 years in the future.
At this point, Ferguson apparently became concerned
about lack of public support for the research programs
and urged the President to make the plans public.

Finally CSSN's work was released to the media. A
public information campaign was launched, alerting
everyone to the dangers of the cosmic ray blast while
minimizing the consequences. Ferguson and other
CSSN members went on the air to urge support for
expensive legislative programs. But contingency plans
for the "survival of the fewest" were kept top secret.

About this time, an international group of Nobel
laureates pleaded for a world united against the
unknown disasters ahead. They advocated spending
more resources on the colonization of space and less on
"ostrich" policies. But their ideas were generally dis-
missed as impractical; there simply wasn't enough time
to develop space travel beyond the solar system.

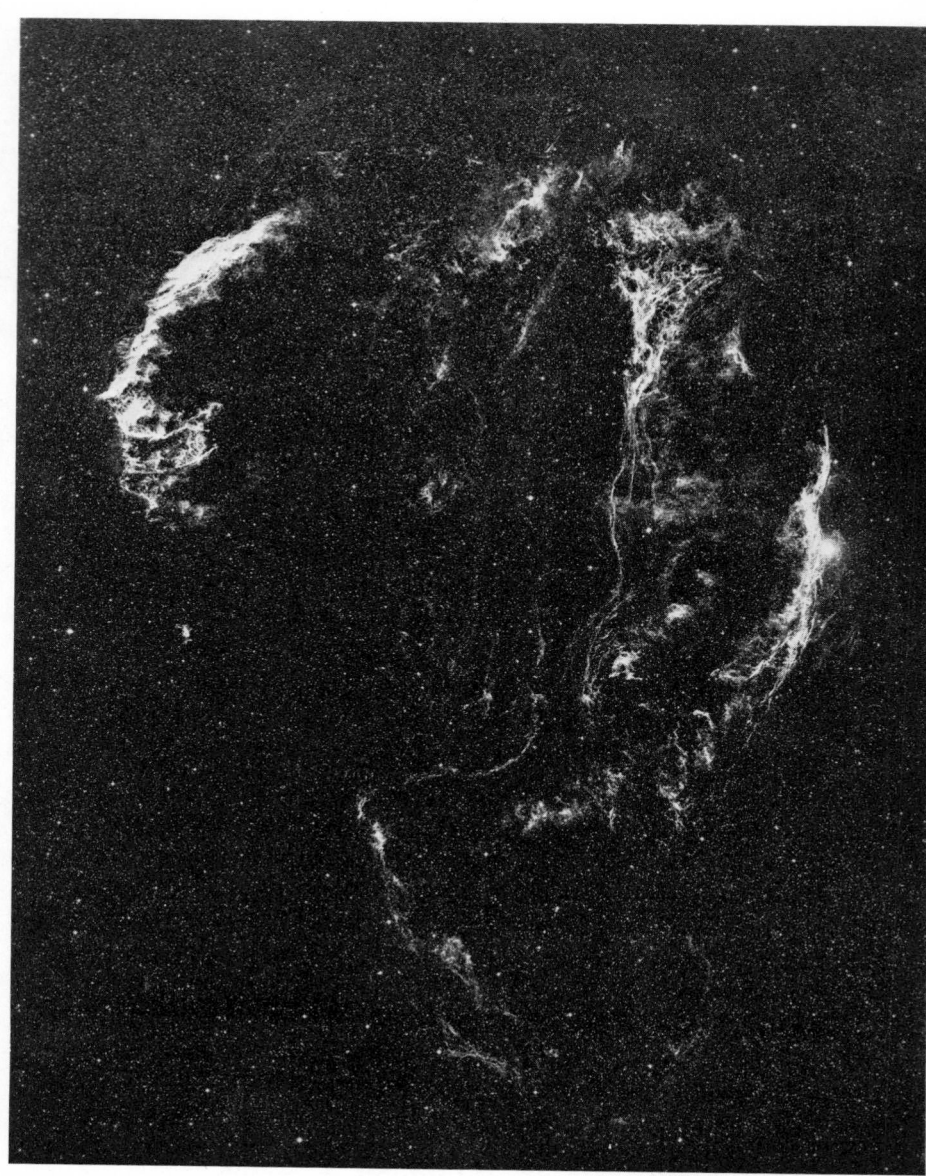

The remains of a supernova that exploded some 50,000 years ago in the constellation of Cygnus. The luminous filaments of matter thread their way between the stars and continue to shine, even long after the explosion, because energy is fed into the nebula from the remaining core of the star that exploded. Each of the tiny dots in the photo is a star in our galaxy. Hale Observatories photo.

The national research programs were funded, probes were launched into outer space by the United States and the USSR, and emergency preparedness was taught all over the world. But public concern soon died down again, as people grew tired of waiting. As it turned out, they had to wait another 39 years. A whole new generation had grown up by then. Ferguson, now an old man by earth standards, still stumped the country warning people of the dangers to come, but nearly everyone was tired of hearing the same old story. The malls were built, weren't they, and no one needed them. Probably a waste of money. Night time was still brighter than a full moon condition, but the stars were once again visible. Everything was back to normal.

To add to the complacency, a report from a space probe that cosmic ray levels had increased in the outer parts of the solar system turned out to be a false alarm.

In July of 2017 three of the deep space probes started to measure a steady increase in the cosmic ray levels. The increases were much greater than expected. Earth had only one day before the cosmic rays reached the monitoring station on Phobos, of Mars. Modified plans A and C were set into motion, and when the cosmic ray counters on earth began to click at higher rates attention turned to the satellites that monitored the ozone layer.

In several months the ozone level dropped 5 percent, and then the rate of change became even more rapid. Atmospheric scientists amended their calculations and found that a complex interaction between various layers of the atmosphere would, in three or four years, bring the ozone level down 40 percent. At this alarming rate, the ozone would be completely gone in 10 years, not enough time to replenish it artificially. A few scientists believed the planet had some kind of natural capacity for restoring itself, but there was no discernible force in operation at that time. All thoughts turned to protection. As a simple first precaution, people were warned not to venture out into the sunlight for more than a few minutes at a time. Many towns in North America had by this time built enormous closed in city-sized malls that allowed continuous shielding from the ultraviolet light of the sun. For a decade many cities had been boasting that they could function adequately when sealed off from the outside air. Some included immense warehouses. Others were giant greenhouses whose solar energy was highly developed. Many communities claimed to be self-sufficient at the subsistence level. In other parts of the world, however, the populace was not so well organized.

In the less-developed countries, and especially those where the star could not be seen, much less notice had been taken of the impending danger. In the first year, deaths by sunburn and skin cancer rose in all the countries where mass communication was less highly developed than in North America and United Europe.

By that time, those chosen for the deep mountain shelters had been sealed off from the outside world. The existence of the shelters and the lists of the chosen few were understandably the best-kept secret of the century. Not even Ferguson had opposed secrecy in this area.

There were three of these shelters, each with a community of 1500 men, women, and children. Every person had been carefully selected for expertise in a needed field, emotional stability, and freedom from dis-ease. Most of the adults were scientists, of course, but there were also artists, poets, musicians, and persons with natural political talent.

With huge libraries on microfilm (the source of most of our information), solar energy plants, air purifi-cation systems, and mining equipment for obtaining raw materials, they were prepared to survive for several centuries. Eventually, they hoped to be able to transmit the core of their culture to a new and safe world.

From the beginning, these watchers from the mountains monitored the rest of the world, recording the last years of their race. (Now we find their records invaluable.) The news that came in from the outside world was grim in the first century. Figures were necessarily incomplete, but world population apparently dropped from four billion to only 100 million in the first 50 years.

The mass horror was never adequately described. Some countries had so few survivors that no one will ever know what really happened. In North America and United Europe tales of disaster came from a number of communities that suffered riots and complete social breakdowns. Some of these are well documented in the archives. Earthquakes, quite normal ones, added to the chaos. One wiped out the city called San Francisco, together with all its carefully planned, self-sufficient structures When people ran out of the ruined buildings, they died from lethal doses of ultraviolet radiation: Millions died horribly, not only in San Francisco but also in other centers of population throughout North America. We plan a special video show on some of these events for later broadcasting.

The worst disasters occurred in low-lying countries. Millions fled the coastal areas as the sea level rose six inches a year. We believe total extinction would have resulted at that time if the fleeing millions had been able to organize and move enmasse to the higher country. Civil wars would have become rampant. As it was, people simply swarmed inland at random, and most were killed by ultraviolet rays or contracted one of the many new diseases caused by mutating viruses.

The planned mall communities, far from the coastal areas, with their air purification and recirculation plants, lasted a little longer, some as much as 100 years, but when they ran out of replacement parts for their machinery, they died out.

The mountain communities survived the longest, about 300 years. When we broke into these shelters eight years ago, the air purification and recirculation systems were still functioning perfectly, but there were no survivors. Probably they succumbed to one of the mutant viruses that plagued earth throughout this period. It is sad to realize that we missed contact with this intelligent civilization by only 200 years.

We have found evidence of some planning for the colonization of the planet Venus. Subtle changes in the atmosphere of both Venus and Mars were produced by the cosmic ray blast and led some scientists to suggest that the atmosphere of Venus might change so as to make that planet habitable. But no colonizing missions were actually sent, so far as we can tell.

The colonies on Mars and the moon were no better off than those on earth. The inhabitants knew they were cut off forever once the great famine hit earth, and by the year 2100 all communication from those bases had ceased. Either they had run out of spare parts for their radio equipment or they had perished. We may never know the deprivation they suffered as they neared their end, but research teams are working now to learn the facts. Our sociologists might like to prepare a special video program on these desolate colonies.

The archives are still being explored to extend our knowledge of terrestrial civilization. We would like to bring to the attention of the Council that solving the language code clearly led to our present ability to explore the Terrestrial Microfilm Archives, but the lack of further funding precludes complete investigation of this material. We trust that the Council will look favorably on the establishment of a permanent Historical Center on this deserted but fruitful planet.

STARBIRTH TO STARDEATH

CHAPTER TWO

The prospect of a supernova exploding within 30 light years of earth is both dramatic and frightening. The effect on our ozone layer, on which our life depends, would be devastating. It is quite possible, however, that supernovae have exploded dangerously near the earth on dozens of occasions during the past five billion years. Could it happen again? Are we really in danger? If a supernova should occur close to the earth in the future, all surface life would be severely threatened, if not wiped out, judging by what we now know about these exploding stars.

Let's trace the evolution of stars from their birth to their death as supernovae. Only by knowing as much as possible about the life and history of nearby stars can scientists predict which ones will someday explode. And to understand our neighboring stars, we really need to trace their story back to the very beginning of the universe.

Events that took place in the early universe had a direct bearing on cosmic evolution and, in fact, on life here on earth, for in those early minutes of the universe hydrogen gas was created, and some of it was turned into helium. Fortunately for us not all of it was converted to helium, or there never would have been stars like the sun, and we wouldn't be here now.

Beyond the stars lie billions of other galaxies. In this photo we see a
cluster of these island universes in the constellation of Hercules.
What creatures live out there? How many civilizations have been
born and died in those galaxies as the result of cosmic catastrophes
similar to those that will test life on earth? Hale Observatories photo.

The first piece of evidence concerning the birth of the universe comes from the examination of light from distant galaxies. Galaxies are enormous conglomerations of hundreds of billions of stars. Our own Milky Way is a galaxy. Light from distant galaxies shows that they are all moving away from each other. When we compare their speeds with their distances, we find that the most distant galaxies are moving away with the greatest speed. Since all galaxies are apparently moving away from each other, the universe must be expanding.

Given that the universe is expanding, it is natural to ask when it was that all things were in one place. Astronomers have found that the expansion started between 10 billion and 19 billion years ago, depending on what assumptions are made about how the expansion has changed during those billions of years.

If there was a beginning to our universe that many years ago, it must have looked very different then. Can astronomers observe any differences? The answer is that they can, because it is possible to look backward in time; the farther outward you look, the farther back in time you are seeing. Light does not travel instantaneously from one place to the next; therefore the farther away an object is, the longer its light has been traveling to get to us. (Light travels at 300,000 kilometers, or 186,000 miles, per second. Astronomers use the *light year* as a convenient standard for distance. It is the distance light travels in one year, about 9.7 trillion kilometers.) If we see an object located one billion light years away, then we are seeing it the way it was one billion years ago.

As we study more and more distant galaxies, we are looking at increasingly "old" light. Look far enough and you see things in a universe that was much smaller, because things were much closer together than they now are. In fact, if you look far enough into space (which means looking far enough back in time), it is possible to see back to a time when the universe was just created. This is precisely what astronomers believe they are capable of doing.

Astronomers now can describe the universe as far back as a fraction of a second after the moment of creation. No one can describe what conditions were like before that, since there was no time then. The known laws of physics break down at the moment of creation; one could say that all the laws of physics came into being at the instant the univesse started its expansion. If there was anything "before" that time, it cannot be described using our known physics. The best description cosmologists have been able to give of conditions before the beginning is that all was "chaos."

This initial moment, the time and place at which it all began, is called the *big bang*. At that instant of definite beginning, all the universe was in one place in time and space. There was no solid matter. Everything was in the form of radiation, and its temperature was trillions of degrees. The big bang was an incomprehensible explosion that gave birth to the universe we now know.

Within seconds the universe was flying apart at the speed of light. In the first few minutes of the universe, the basic units of atoms; the protons and electrons, were created. Since a hydrogen atom consists only of one proton and one electron in orbit about it, we can say that in the beginning hydrogen was created, even though the electrons couldn't stay in their usual orbits around the protons at such high temperatures. After approximately 15 minutes, neutrons, the third kind of basic particle, were formed. These events in the early minutes determined whether the universe would evolve in a way that would set the scene for the ultimate emergence of life.

All the while, the universe continued to expand and, when its temperature dropped to a few hundred million degrees, the next important step was reached. At these temperatures, the hydrogen (actually the protons, one of the basic particles of which atoms are made) combined with neutrons (another basic particle) to form helium in a process called fusion. (When the temperature is too high, the atoms will remain separated into their basic particles.) Fusion proceeds as long as the temperature is high enough. This is the process that generates energy inside the sun, other stars, and in hydrogen bombs.

As the universe cooled, fusion of hydrogen produced helium, but fortunately for us the universe cooled fast enough so that its temperature quickly dropped below the 10 million degrees needed for fusion to occur. If it hadn't cooled so fast, all the hydrogen would have become helium. In that case we would not be here today, for there would have been no hydrogen left to form stars and no stars to form the more complex atoms needed to make planets and life.

The mating of hydrogen nuclei to form helium therefore ceased as the universe continued to expand and cool. The universe still continues to expand and cool, even today, and its overall temperature is now only three degrees above absolute zero. (Absolute zero is believed to be the lowest temperature anything can ever reach, a temperature at which all motion within atoms ceases. It is −273.4° Celsius.)

After the universe was created, a staggering amount of matter hurtled outward into surrounding nothingness. It moved in all directions at the speed of light and, somehow, large clouds containing enormous amounts of this primeval matter, amounts equivalent to a hundred million suns, gathered and started to coalesce. It is not known exactly what triggered this process. Perhaps irregularities in the early universe acted as points of attraction for the clouds from which galaxies like our own Milky Way subsequently grew. The sun is but one star out of hundreds of billions of similar stars in the Milky Way galaxy. And it is believed that there are hundreds of billions of other galaxies spread throughout the universe, each of them filled with hundreds of billions of stars and who knows how many planets. On at least one of these planets life emerged and endured all the way to intelligence.

In the galactic clouds matter coalesced further, and smaller clouds separated out from the whole. The amounts of matter in each of these entities was enough to form hundreds or even thousands of stars at a time, and when the moment of birth was ripe, the space inside the clouds began to glow as hundreds of stars began to shine. What process was at work here? Very briefly (for we will describe the birth of a star in greater detail in Chapter 4), the matter swirling about inside the galactic clouds continued to condense as two immense forces struggled against each other. Gravity pulled each cloud together, while the collisions of all the particles in the cloud generated heat that tended to force it apart. At a critical point, the contracting cloud broke into even smaller entities that developed into individual stars. Each of these continued to contract, and, in doing so, became hotter and hotter because the particles in it were rushing against each other ever more violently.

When the internal temperature of a star cloud reached a staggering 10 to 15 million degrees, the hydrogen atoms started to join with each other to form helium. In this fusion process enormous additional heat developed, and the center of the cloud became a generator of energy that radiated out into space. The star cloud began to shine. Now the fusion process acted like a furnace, keeping itself lit by burning more hydrogen. This newly generated heat just balanced the inward pull of gravity, the cloud contracted no further, and a star was born.

What happens to a star after that? Modern technology and a broad knowledge of the laws of physics allow astrophysicists to make computer models of a star's life and death. Simulating the evolution of a hypothetical star greatly speeds up the process! Instead of waiting 50 million years to watch a star evolve and die, astrophysicists can run computer programs for mere hours and watch what the equations reveal. These equations tell them that as a large star matures, its interior gets hotter and hotter, and elements fuse to form heavier elements. In this way more and more of the basic building blocks needed to form planets are made inside the star. Some stars, particularly those containing more than three times the mass of the sun, ultimately make iron. At this point a new phase in the star's evolution is reached.

Surprisingly, iron requires even more heat to play out its role. When the center of the star reaches 600 million degrees, the iron literally sucks in the surrounding heat of the

A star cluster known as NGC 2682 in the constellation of Cancer,
photographed through the 200-inch telescope of the Hale
Observatories. Stars are generally born in large numbers from one
cloud of original matter. Such a star cluster can exist for many
hundreds of millions of years before the individual stars wander off
into surrounding space. The sun must have formed as a member of
such a cluster nearly five billion years ago, but all the other members
have long since gone their own way. Hale Observatories photo.

star's core and acts like a giant fire extinguisher. Suddenly, the core cools down, and the fire at the center goes out in minutes. The matter at the center of the star now implodes violently, because gravity can act without having to struggle against heat pushing outward. There is nothing to prevent a violent implosion.

As the core collapses, all its matter is crushed so tightly that the protons and electrons (the third of the trio of fundamental building blocks of matter) combine to form neutrons; the core becomes a solid ball of neutrons. An object made only of neutrons is called a neutron star. Its density is incomprehensible to us. A teaspoonful of neutron star would weigh a billion tons!

In some stars the collapse is so violent that even the neutrons interact and combine with one another. This creates such an enormous force of gravity that nothing can stop the implosion. It continues until the center of the star becomes a black hole. Nothing, not even light, can ever escape once it comes within the greedy reach of a black hole. However, only a few of the largest stars are expected to become black holes, and these will be discussed later.

While the core collapses, the outer layers of the star are still generating heat, momentarily oblivious to the cataclysm occurring at the center of the star. The implosion occurs within minutes, and the overlying layers of the star find themselves hanging up there with nothing to hold them up against gravity. The consequences are disastrous. The outer shell of the star collapses inward and falls onto the neutron star just created at the center. This infalling matter actually smashes onto the neutron star and bounces off again in an unimaginably violent explosion.

The temperature of the exploding gas reaches billions of degrees in a supernova blast that totally shatters the remains of the star. It hurls matter outward at 20,000 kilometers per second. The flash produced by an explosion like this can generate as much light as a billion stars and can be seen even in distant galaxies. A supernova may produce as much light as the whole of its parent galaxy. If any of the stars visible in our night sky were to become a supernova, we would have no night, for the star would shine at least as brightly as the full moon.

Life on any planets in orbit around a supernova would immediately perish, so it is reasonable to ask whether our sun might someday become a supernova and exterminate us all. The answer is no. The sun isn't large enough to explode in this way. The type of stars that become supernovae never have time to grow old; their age is usually not more than a few hundred million years — too short a time for the emergence of life around them. Scientists believe that life typically requires billions of years to emerge on a planet, and only near safe, old stars like the sun can we expect to find good places to live. So we are probably in no danger of this kind from our own star.

The sun will die someday, but its death will be considerably less violent than that of a supernova. After it shines for another five billion years or so, it will enlarge for a while, then fade away and die as a planetary nebula. This is the name given to beautiful rings of matter that are ejected fairly gently (from a cosmic perspective) by stars like the sun in their old age.

In fact, and fortunately for us, most stars die in this quiet way; otherwise we would be blasted by a lot more supernovae. About once every two years in our galaxy some star, somewhere, becomes a planetary nebula. On the other hand, a typical galaxy should experience a supernova only once every 50 years.

Despite this estimate of supernova frequency there has been none in the Milky Way since the seventeenth century. The next one is long overdue. Where will it occur, and will it harm life on earth? No one knows. In fact, no one knows with certainty which type of star actually explodes to become a supernova, because no one has ever seen the "before" and "after" star. All the "local" supernovae in our galaxy shone very brightly and could be clearly seen from earth, but they occurred before the telescope was invented. Supernovae occur in distant galaxies, but they are so far away that astronomers

cannot see individual stars before they explode as supernovae.

In order to understand a supernova better, astronomers
monitor any change in its brightness or in the characteristics
of its light. In particular, they study the spectrum of the light.
A spectrum shows dark and bright lines on a piece of film
that represents the way the intensity of the light varies with
wavelength (or color), ranging from red to blue light. The
lines on the spectrum can even indicate the presence of specific elements and their relative abundance in that distant
star. A change in the spectrum alerts scientists to what kinds
of changes are going on in the star.

In recorded history supernovae have been seen several
times, but none of them exploded very close to our sun.
The nearest, seen in 1054, occurred about 6000 light years
away; it shone so brightly that for months it could be seen
during the daytime. It was still visible years later as a bright
star in the night sky. Now, 900 years after the event, we find
a luminous filamentary cloud of gas, called the *Crab nebula*,
at the position of that supernova explosion. The Crab nebula
is the shattered remnant of the stellar cataclysm of 1054. At
its center a neutron star spins rapidly, emitting pulses of
light, radio signals, and x-rays. This pulsar acts like a gigantic
cosmic lighthouse warning us of the dangers that lie in wait
inside the nebula. It will spin for millions of years as the nebula continues to expand. Finally the nebula will fade from
view, and there will only be the pulsar to indicate to future
astronomers that there was once a star there.

Countless planets today must be orbiting stars that are
very near the site of a recent supernova explosion. Their skies
must be filled with wreathes of beautiful, wispy filaments
that are, in turn, filled with lethal cosmic rays. The sight
would be amazing to behold, preferably from a great distance, since it would be fatal to experience.

A cosmic ray is a small particle, such as an electron or a
proton, that travels through space at speeds approaching that
of light. Cosmic rays are produced in large quantities during
supernovae explosions as well as during novae (small stellar

The Crab nebula, the remains of the star that was seen to explode in
A.D. 1054. Now, some 900 years later, the remains of that explosion are
still hurtling outward at enormous speeds, sweeping past hundreds
of other stars and their planets. Life on any of those planets would
be severely damaged, if not totally destroyed. Hale Observatories
photo.

explosions), flare star explosions (explosions on the surface of certain stars), and in explosions on the sun. They continue to be produced by not-so-dead neutron stars long after a super- nova explosion is over, and this makes supernovae the prime source of these particles.

Matter in the shell ejected by an exploding supernova is extremely hot. It is also ionized; that is, the basic atoms have lost some of their parts, especially the electrons that orbit about their nuclei.

As the shell of ionized (very hot) matter expands out- ward, it encounters and sweeps up interstellar matter that is already floating between the stars. This includes gas and dust particles and magnetic fields threading their ethereal way through interstellar space.

A magnetic field is a region of space in which magnetic forces can be experienced. All magnetic fields are produced by moving electrons. In much the same way, a magnetic field is set up near a wire carrying electrical current; this is often demonstrated in simple school lab demonstrations.

As the supernova shell expands into surrounding space it continues to pile up magnetic fields and gas ahead of it. It forms what is called a shock front in which the field can be- come stronger. In addition, the star itself had an internal magnetic field, and when the star shattered this field was dragged about into the filamentary patterns that show up in supernova remains (see photograph of the Crab nebula). The result is that some of the ionized particles that originated in the exploding star get swept along by the magnetic fields, ac- celerating until they travel at nearly the speed of light. They are then called *cosmic rays*. (A side effect of this is that the cosmic ray electrons also spiral about the magnetic fields. As they do they radiate away some of their energy in the form of light and radio signals and this is why we can see the super- nova shells shining.) Some of the cosmic rays leak out of the nebula and wander off amongst the stars. In fact, cosmic rays from the remains of supernova explosions fill all of space in our galaxy.

Most of the cosmic rays that ultimately reach earth originated in the dying gasps of distant stars hundreds of millions of years ago, and they have been traveling through space ever since, aided on their journey by the magnetic fields that wend their way between the stars. As they spiral about the interstellar magnetic fields, they cause the whole Milky Way to radiate strong radio signals that are easily picked up here on earth.

When cosmic rays reach the earth, they can affect us in two very different ways. First, they can completely shatter molecules in the air above us (provided they make direct hits) and thus have a serious effect on our ozone layer. Second, some cosmic rays can penetrate the atmosphere and pass right through our bodies.

It appears that large doses of cosmic rays would affect the ozone layer that surrounds our planet well before the cosmic rays themselves could affect us. (Ozone is a form of oxygen in which three oxygen atoms combine to make a stable molecule. It is different from the oxygen molecules we breathe, which are made up of pairs of oxygen atoms.) The earth's ozone, in a layer about 30 kilometers above the surface, acts as a vital shield to the lethal ultraviolet (UV) radiation that is directed toward us by the sun. If the ozone layer were removed, the solar ultraviolet rays could wipe out a significant fraction of life on our planet in a fairly short time.

Let's consider the mechanism by which cosmic rays (mostly protons and electrons) could deplete our ozone layer. When the protons in cosmic rays collide with molecules in the earth's atmosphere, they shatter them and produce an enormous number of electrons. These electrons are stripped out of their orbits in the atoms and molecules of the atmosphere. The electrons, in turn, strike nitrogen molecules, each of which consists of two nitrogen atoms. In many cases, one of the nitrogen atoms may also have an electron stripped out of its orbit (ionized). These ionized nitrogen atoms then combine with oxygen molecules to produce oxygen atoms and nitric oxide (NO) within the ozone layer. This is where the trouble starts, since the NO now combines with the ozone (O_3) to form nitrous oxide (NO_2) and oxygen, thus destroying the ozone. This nitrous oxide, in turn, combines with oxygen atoms again, to form nitric oxide plus oxygen molecules. We are therefore left with the same amount of NO with which we started, and this can again be used to destroy more ozone!

Cosmic rays of especially high energy can penetrate the ozone layer even now. If the ozone layer were destroyed, there would be nothing to protect us from the sun's ultraviolet radiation.

Cosmic rays are really a form of radioactivity. Because of their great speed and energy they can damage living tissue. Should a cosmic ray directly strike a molecule or a cell in your body, it could shatter that cell and produce damage we could regard as a mutation. In general, the mutation of a single human cell will not change the course of anyone's life, but cosmic rays are believed to be capable of changing the genetic makeup of an organism. In fact, it is likely that, during the evolution of life on earth, cosmic rays have played an important role in mutation of all living things, especially of the simplest organisms. The simple viruses floating in the air about us could suffer massive mutations if they were struck full force by cosmic rays.

Large doses of cosmic rays are therefore not conducive to health. Indeed, direct blasts of cosmic rays from nearby supernovae have in the past been blamed for the extinction of species on earth, but such destruction of living things would require a very nearby supernova only a few light years away. This is an extremely unlikely event, so unlikely that it cannot in itself account for the extinction of species. However, cosmic ray doses larger than normal may, indirectly, through their impact on the ozone layer, produce a devastating effect on life.

Simple calculations suggest that a supernova sufficiently close (30 light years away) to remove half of our ozone layer should occur about every 200 million years. This means that the earth should have been zapped by destructive cosmic ray doses dozens of times since it formed nearly five billion years ago. Viewed in this way, it is surprising that any life has survived here at all. It is clear that the path taken by evolution-ary forces on earth must have been intimately affected by rel-atively nearby supernovae.

If a supernova is to produce a near-fatal effect on earth, the exploding star must be located closer than about 30 light years away. If it were located 50 light years away, only those species most vulnerable to small changes in ultraviolet doses would suffer, because the ozone layer would undergo only a small decrease and nature would probably find some way to counteract the decrease. The surprising thing is that even though cosmic rays travel at nearly the speed of light, they would take a long time, perhaps even hundreds of years, to reach us after we see the explosion. This is because magnetic fields in space deflect the cosmic rays, causing them to wan-der through space somewhat haphazardly; they never travel in straight lines. The only way we could find out how long the cosmic rays from an exploding star 30 light years away would take to get to us, would be to do the experiment!

Some scientists have suggested that nearby supernovae may nearly extinguish life on most planets about once every half billion years. The fact that we are here at all might there-fore mean that we have just been very lucky. Perhaps, if earth had suffered the fate of a typical planet, we wouldn't be here. It is always possible that some planetary systems have re-ceived more than their fair share of cosmic ray blasts, and we, and perhaps a few others, have been lucky and missed out on our share for several billion years. This sort of specu-lation is relevant when estimates are made of the number of inhabited planets in the Milky Way.

It seems virtually certain that a nearby supernova occur-ring every few hundred million years would be enough to hamper evolution on earth, so that life would be prevented from reaching our stage of development. But since we are here, we must conclude that if supernova blasts did hamper evolution, they didn't stop it completely. However, such blasts might explain why it took so long (nearly five billion years) for us to come on the scene! Unfortunately, we know of no other inhabited planets with which to compare our earth, so we don't know if we are unique or not, whether we are particularly advanced or not, or even whether we are the only ones who made it this far.

If we have been lucky, we should ask how much longer our luck can last. Could we be blasted any day? The answer is that we might be, but no one can call the odds with any certainty. Our day could come this year, or 10 thousand years, or a million years from now.

THE GOOD HOPE
CHAPTER THREE

Barek paused, one gloved hand on the handle of the air-lock. He didn't particularly want to go back inside the starship *Good Hope*. The more time he spent out here, the more confined he felt in there, and the constant discipline was wearing after a taste of freedom. He looked back once more at the intoxicating vista of blazing stars silhouetting the immense sail that had propelled him from one of them to another all his life.

Out here in the soft darkness he felt emotions unlike any he had known before his first extravehicular assignment. And the novelty never wore off. Out here it was easier to dream of being on some planet, perhaps within a year. He tried to imagine for the thousandth time how it would feel to be held down permanently by gravity. At least he'd be free to be outside with the stars whenever he wanted!

He'd been outside the ship for three hours now, supervising the crews sent out to repair the sail after their ill-advised approach to F 25-32. It had taken four months to make the repairs, and the crews had worked reasonably hard. They all knew time and fuel were running out and that the mission of everyone on the *Good Hope* was to find a suitable planet for colonization before it was too late. Too late for them, and for all those left back on earth so long ago.

This was the last day of repairs, and in a few minutes everyone would have to go back to the tedium and tight discipline of the ship world. He hoped it wouldn't be much more than a year inside.

The others, floating easily along the outer shell of the ship in their metallic suits, seemed more relaxed than usual to Barek's experienced eye. A little slower, and looser. They were still staying together in the assigned pairs and small groups that safety on EVAs demanded, but their pace was down, and sometimes they stopped entirely for no other reason, Barek supposed, than wanting to immerse themselves in the quiet and glory of deep space just as he had been doing. It was quitting time, but there was no point to bringing them in right away. Let them stay out late. Acceleration would soon enough end this kind of freedom.

Barek stopped, raised the clasp, and entered the airlock. Time to make his last report on the condition of the sail. The new hydrogen collectors were secured, and the crew was already making its final check on the action of the giant antennas that protruded like rude growths from the hull.

He was just removing his helmet when Kara floated into the suiting room. "Thought you were never coming back in!" she snapped, tossing her head. Her angry expression painted a sharp contrast to her leisure tunic, embroidered with birds of paradise. "Everyone's been waiting, but I guess you don't care about that!"

Barek's mood of mystical awe evaporated under this onslaught, but he tried to answer patiently. "Kara, love, just relax. There are more important things than split-second shift changes. You're off duty, now aren't you?'' Hopefully, he traced the length of her spine with the back of one finger. "I'll see you as soon as I finish my report."

She squirmed away. "Stop it, Barek. Don't play games now. Captain Dolvar wants to see you right away."

Barek immediately pushed off down the corridor, though the initial impulse would help him only as far as the turn in the narrow passageway. After that he had to use the hand grips attached to the white-painted walls. He wished Kara had told him sooner about the captain's summons. Although he often chafed under discipline, he supported it because he knew how necessary it was. And he had to admit that the captain was fair in his dealings with the crew and tried to find out what was on everyone's mind.

On the control deck, the captain stared silently at the view screen, an electronic representation of the light picked up by the *Good Hope*'s antennas. Of course, the view wasn't as dramatic as it was outside, where the crew was now checking the radio telescope. No color, for one thing. He wished he could still go on those EVAs. The repair crews didn't know how lucky they were. Or maybe they did, he thought ruefully; they seemed to be staying out late just to look at the pretty stars. That was going to change right away.

Barek's smart knock on the bulkhead brought Dolvar out of his reverie, and he turned as the young officer floated into the room and saluted.

"Barek reporting, sir."

"Right. Why isn't the crew in yet? You told me all repairs would be finished by 20:35, but everyone's still out there. What's going on?"

"Well, sir, this is their last day out for a long time, and I thought. . . ."

"You thought! I thought, too, Mister. And one of the things I've been thinking is that we should have accelerated half an hour ago in order to set optimum course for binary D 36–39 by noon tomorrow. So that we can arrive at D 40–31 only one planet-year behind schedule." He lowered his voice dangerously. "You realize, perhaps, that we can't accelerate until your people are back inside the ship?"

"Yes, Captain, but. . . ."

"And perhaps you have considered the total red accumulation since these repairs began? Do you think so little of 350 rads that you can expose your people to even *one minute more* of those cosmic rays than is absolutely necessary? Has it occurred to you that every member of those repair crews is taking a chance on radiation sickness?

"Yes, Captain, but. . . ."

"Don't yes captain me," Dolvar thundered. "You've got six minutes, Mister!"

Barek turned and propelled himself back along the corridor as fast as near-zero gravity would allow. Dolvar watched him leave, then turned to Nadra, a serene, black-haired woman who monitored the computer on the other side of the control deck. Unlike the rest of the crew, whose marriages has been arranged, he and Nadra had chosen each other freely. At times like this, he appreciated her quietness and calm intelligence.

"Kind of rough on him, wasn't I?" he muttered.

"Yes, but sometimes it's necessary," she replied. "We're running 45 minutes behind schedule already. I'll have to reset our course now."

"Still, I think I'm showing the strain," Dolvar said as Nadra turned back to her computer terminal. In fact, the entire ship, officers and crew alike, were feeling the strain of resuming the search for a livable planet.

He remembered how it had been when the sail was damaged. They had been approaching a now-distant planetary system, slowed from more than half the speed of light to only 100,000 kilometers an hour—certainly a safe speed. But they had run into an unexpected assault of asteroidlike rocks in orbits that encompassed the entire planetary space. There was no way to avoid them, but they didn't know that at first. By the time they had tacked and explored several possible avenues, the sail and two of the collectors had been extensively damaged. Furthermore, remote scanners told them that no life could inhabit any of the 12 planets anyway. The whole venture had been a waste.

They had turned and accelerated out to 350 billion kilometers from the asteroid orbits, where they could safely send out the repair crews. Dolvar still grimly recalled the mood of the crew. They acted as if they were on holiday, they were so delighted to be released from the confines of the ship. But he had to admit that Barek hadn't let the crew slack off, at least not until today.

"All right, Dolvar," Nadra told him. "We're all set as soon as Barek checks in." She indicated the course and systems data flashing across her computer readout screen. Tacking into the stellar wind was a job for the helmsman, but her computer plotted the macrocourse and later, when they were under way, provided a constant check on their stellar flight.

At the moment, the wind was very low and the sail wasn't providing much help. The normal interstellar cosmic ray stream, coming randomly from all directions, could not propel them, and without some speed the collectors were useless, too. Their initial acceleration would depend entirely on their dwindling hydrogen supply. They had already used too much of it on the ill-fated venture into the 12-planet system. The enormous sail,

hundreds of square kilometers in area, could catch
a star's wind and accelerate them to velocities of many
thousands of kilometers per second. Then enough
hydrogen would be funneled into the compressors to
propel the starship to stellar escape velocity without any
auxiliary power drain, but they had to start fairly close
to a star to do that. The fusion power then would get
them up to 50 percent of the speed of light, their usual
cruising speed after leaving a star. But right now they
had to get to the nearest star to make use of its wind.

Everything was quiet on the control deck until the
viewscreen showed that all the crew members were
finally back inside the *Good Hope* and Barek's voice
came over the intercom. "All hands aboard, Captain."

"Roger." The captain pushed the buttons that
brought in the auxiliary power. "We're flying," he told
Nadra in the ritual phrase, and watched appreciatively
as she nodded her lovely, sleek head.

"Good," she said. "Let's grab a cup of caffe."

In the officers' lounge, they found a small group
already reclining around the caffe table, most of them
dressed in embroidered leisure tunics. Two years earlier, embroidery had been a fad for several months, for
both men and women. Dolvar's grandmother, in the
first shipboard generation, had handed down the skill,
and with each revival it became more elaborate.

The foursome around the low table had been talking in low tones, but they fell silent as Nadra and the
captain approached.

He knew why. The binary, which beckoned with
awesome beauty and the promise of badly needed
power, also held risks. The *Good Hope* had never been
as close to a binary as they would have to go. For its
whole history it had bounced off one single star after
another. No one had observed the interactions between
this particular red giant and its dwarf companion until a

few months ago. The crew, he realized, had cold feet, but the *Good Hope* was going in anyway. And it was high time to get the talk out in the open.

"Well, we're flying," he announced as he threw himself down and hooked onto a pod.

"We know," said Marret, a petite redhead who served as ship's astronomer. Her quick smile failed to hide her tension. "Just think, getting within a tenth of a light year of D 36–39! No doubt we'll learn a lot about binaries we never knew before."

"No doubt," repeated a junior officer. "But will it be worth it to us? Anything could happen."

"Not likely. I feel quite confident about going in," Dolvar answered. "It looks perfectly stable from here, and you know how badly we need the power boost. The giant should get us to point seven cee. Besides, Marret is dying to analyze what's really happening in a binary." He grinned at her, then continued seriously. "I don't see any alternative. Either we get a power assist from D 36–39, or we spend the next four years out here in midspace, moving about one-tenth cee. We can't afford that, and neither can anyone on earth.

"That's true, Captain," said Marret, "but we can't be sure it's safe. Of course I'm scared, but it just might be a prudent fear." She squeezed the last drops out of her "cup" and look squarely at Dolvar. "We all know the red giant has a good wind all the way across the Roche Lobe, but what if it blows off enough stuff to explode on the dwarf? That could happen any time, and it would spell disaster! You seem to forget that we wouldn't see it until long after it happened — and then it would be too late."

"Look, Marret, I know all that. But we've gone over the possibilities a dozen times. Our data banks show nothing unusual on this pair. So what chance is there that they'll kick up in the next year? There's no recur-

Scenic Overlook. Etching. An alien star looms over a surreal landscape.

rent nova there, nor has there ever been a nova as far as
we can tell. It's just one of millions of harmless binaries
in the galaxy. It's a bit of a chance, sure—but a very
minor chance in my opinion. And we'll continue to
transmit all computer functions back to the *Gaia*. She's
gained some on us these past months, but she's still ten
years behind, and if we run into trouble, she'll have
time to avoid the same trap. In other words, if we don't
make it, she will. I'm optimistic. We'll go in, get our
boost, and blow out of there."

"That's right," said Barek, who had just come in.
"Why worry?"

Marret still looked unhappy. Finally she said, "We
do have a few months of monitoring before we get too
close to change course. But I have bad vibes about this
binary system." She laughed unsteadily. "Guess I'm just
getting nervous. How about another round of caffe
before we turn in? Who wants some?"

The next day meant the resumption of interstellar
routine. Work cycles were repetitive and time-con-
suming, deliberately made so to keep the crew busy.
Barek and his maintenance crew polished the bulkheads
six times a week, recycled air and water filters twice a
week, tore down the electrical generator, cleaned it, and
put it back together while they switched over to the
standby generator. It was all very boring, and Barek was
glad for diversions.

The best part of his life was his marriage to Kara.
Everyone had to take a mate at sixteen, since the psy-
choshrinks had decreed, three generations ago, that sex
within marriage would be better for health and morale
than any other arrangement. Maybe so. Barek at 16
only knew he wanted to get to that new planet. If he
had to get married, he didn't mind too much. Now, four
years later, he was amazed at his former indifference to
women. Kara was beautiful, sexy, often comforting, sen-

sitive to his moods at certain times and unbelievably insensitive at other times. He never knew what to expect from Kara, and that suited him. She certainly wasn't boring. He was also pleased that they probably wouldn't have to wait until 30 to have children. That was a ship decree to space out generations, but on the new planet—given adequate resources—they might decide to populate as fast as possible.

Another good feature about ship routine was the games. Since they were constantly in zero gravity except during acceleration, which amounted to one one-tenth gee, each member of the crew was required to exercise in the gravity rooms at least once a day. These spinning units, located at the center of the starship, were constantly adapted to facilitate new sports that crew members invented. Barek was looking forward to his next game of ricochette, especially since his team was leading in the playoffs.

As the months ticked by, Dolvar and Nadra spent a lot of time talking about their own future. Their time for children had already come, but they had decided to wait until the colony could be established. Each time they reviewed the data on D 40–31 they grew more hopeful, and anyway, enough children had been born on this ship!

Nadra thought a lot about the *Gaia*, too. Since she wouldn't live long enough to see a message from earth and would never know its fate, she didn't think too much about it, except for replaying some old family tapes once in a long while. The *Gaia* was another matter, for on it were people with whom she could feel some kinship. Messages received from it were more than 10 years old, of course. The woman who had taken over the computer, according to recent messages, seemed like a much younger sister, but they were actually the same age. Funny, to Mavir she would seem the younger one! Perhaps some day they would be friends.

Time passed, and the distant binary star system grew closer. No one on the *Good Hope* had an inkling that this binary was in fact far from inactive. A hundred thousand years before, the more massive member of the binary had undergone a violent upheaval that had sent clouds of matter crashing into the other star. As the gases fell, they had heated violently and exploded, tearing off into space on an unimpeded journey. The cloud had shone brightly, and any astronomers then in the galaxy would have seen it. But at that time there were no astronomers on the planet earth. The same pattern repeated itself 50,000 years later and again about 10,000 years before the starship approached. But no one who saw it was able to record the event.

The red giant constantly belched streams of matter onto its companion dwarf star. Usually these streams, which Dolvar planned to take advantage of, were moderate and harmless. But now the red giant was slowly oscillating again. The oscillations made the surface swell, collapse, and swell again. Deep inside the star forces were at play that humans had never studied. Nuclear transformations created bursts of energy that struggled upward to the surface layers, which heaved and rolled like immense tides.

In her observatory Marret pressed the button that removed the cover of her telescope. She swung the huge eye toward the red star ahead of them. The telescope bent the starlight and focused it into the spectrograph, where she could examine the nature of the light in great detail. A computer scanning system did the analysis quickly, and told her which atoms were emitting the light and how they were moving.

Everything appeared normal at first. Then Marret

noticed that a few of the numbers, displayed on the screen seemed different from the previous day. She touched the screen and changed the analysis program. The new data were stored while she instructed the machine to display a comparison of the data for the two days. She studied them carefully, without hurry. This analysis was too important to be rushed. Thirty minutes later, still dissatisfied, she stored the telescope and pushed out of the observatory, a summary printout tucked under her arm. She needed more time, she thought, to really grasp the implication of these new figures.

Then, moving along the corridor between the observatory and the lounge, she suddenly realized what she should have seen immediately. The data indicated regular and quite unexpected motions in the atmosphere of the giant star. Quickly she propelled herself back to the computer and displayed the data again. Then she returned to the telescope and repeated her measurements.

It was obvious beyond any doubt. The massive star had begun to oscillate and even now might be throwing huge masses of additional matter out into space. It would be dangerous to attempt to sail anywhere close to it. If it acted like other binary stars, then some of that matter could fall onto the smaller star with enough force to produce the enormous explosion she had feared all along.

She called her assistant and told him to continue monitoring both stars while she reported to the captain.

She found him at the control desk. "Captain Dolvar," she told him formally, "we're in serious trouble with D 36–39."

"What's the matter?" he asked. "What's happening out there?"

"Essentially, the transformation of the stable oscil-
lation of the red giant into an unstable mode threatens a
serious explosion any minute. It's *heaving*, Captain!"

"What are the odds?" he inquired. "And how big
an explosion?"

Marret handed him the printout without comment,
and Dolvar scanned it quickly. With only about two
supernovae in the entire galaxy in a century and a planetary nebula every couple of years, he thought they
were hardly likely to come anywhere close to either situation. Looking up, he said, "Even ordinary novae—
and I admit they're common in binary star systems—
give danger signals we aren't getting here."

"I know, Captain," Marret said. She peered over
Nadra's shoulder at the view screen, then wandered to a
pod and hooked on. "But this could get very difficult
for us if that surface breaks loose. I can't give you the
odds yet, but I'd say they are pretty high."

Dolvar studied the printout more closely.
Reluctantly he came to the same conclusion: there was
danger ahead.

"You're right, Marret, but we're too close now to
change course, only one-tenth of a light year away.
Continue to feed all data into the computer, of course,
and put all your analysts on overtime, so we have
24-hour monitoring. We'll try to figure out just what
alternatives we have, if any."

Even as the captain and Marret were talking, the
red giant was hurling trillions of tons of matter down-
ward toward the dwarf star, whose light interacted with
the immense blast of approaching gas. Physical processes, pushing and pulling the atoms, loaded them
with enormous amounts of energy. A critical point was
reached as the matter close to the dwarf became so
highly charged with energy that it could no longer continue its fall. With incredible violence it exploded. By

the standards of a nova the explosion was insignificant, but by human standards it was almost beyond comprehension. The *Good Hope,* of course, had no way of knowing about the explosion until its light reached the ship a month or so later.

After his discussion with Marret, the captain decided to bring the full resources of the ship to bear on the new problem as they continued their steady progress toward the red giant. But as they studied the data, it became obvious there was little they could do if the oscillations proved to be a danger. Changing course at their present speed would require enormous power. Basically, they would have to slow the ship first and then redirect it, and that appeared out of the question, given their dwindling fuel supply. The problem really centered on whether there was immediate danger in the red giant, or whether its oscillations were normal and harmless.

Nadra was on watch duty when the distant stellar explosion lit up the viewscreen. The starship was plowing toward the binary at over 60 percent of the speed of light, heading straight for catastrophe. She sounded the alarm, and within minutes the control desk was a frenzy of activity.

Marret, in her observatory, immediately discounted the possibility that the explosion was a nova, primarily because it was too small. It was closer to the level of explosion that occurred in a flare star. Flare stars had been studied for centuries, but there was little data on what actually happened inside the stars. What was now obvious from the spectrum of the flare was that matter was streaming from the dwarf star in tremendous quantities. This stream would generate a shock wave ahead of it, and the starship was headed directly toward it.

Marret estimated that by the time they reached the shock, the light flash would have totally died down, but the shock wave itself would still be moving out at over 10,000 kilometers per second, a thirtieth of the speed of light. More dangerous to them was the fact that the shock wave would pile matter up ahead of it, compressing it to high densities. When the starship plunged into these quantities of matter, it would risk further damage to the great sail.

The next few days saw a lot of computing and argument among the crew members, as everyone went over the data now pouring in. Surprisingly, their mood was generally optimistic. Dolvar did his best to encourage the optimism, which he himself shared. The ship would probably not be hampered too much by the shock wave itself. It had barely started to travel outward from the dwarf star, and it wouldn't have swept up much interstellar matter by the time it reached the *Good Hope.*

Then, in the midst of the general mood of relief, Marret calculated how strong a wind was likely to be behind the shock wave. The shock itself might go unnoticed, being only a brief event in time, but the wind produced by the gases streaming along behind it would be considerable. The *Good Hope* was about to sail directly into an interstellar hurricane.

Marret alerted the captain of the new danger, and efforts were soon under way to deploy the sail in an orientation that would minimize the effect of the wind. Barring a change of course, which was now totally out of the question, there was little they could do except swing the primary controls for the sail so that it slowly took on a position some 45 degrees to the direction of the expected wind.

Captain Dolvar was at his command post when the sensors told of the approaching shock front. The interstellar density suddenly went up a hundredfold as the ship dived into it. But the shock wave was so narrow

that they were through it in only two days, and during
that time they hardly felt it. Then, as they emerged from
the shock wave, their monitors picked up the wind
blowing behind it. It was strong enough to swing the
hurtling starship almost imperceptibly off its course.
They were now sailing into a wind that the designer of
the craft had never dreamt of, some 8000 kilometers per
second.

As the course of the starship began to shift, the
captain, Nadra, and the helmsman hurriedly planned a
further change in the orientation of the sails. They had
hoped that the buffeting of the wind would merely
swing them outward in their orbit past the binary, but
the fierce flow of particles from the dwarf was too rough
for that. It became increasingly violent and turbulent,
with enormous gusts hitting them irregularly. The sail,
designed for normal stellar winds, began to break loose.

The crew watched on the viewscreens as the outer
quadrant of each of the two main sections of the sail
began to float free of their stays. These outer parts
didn't have much structural support, and the crew had
always known they were the most likely to be damaged.
Within hours the ends of the outer sail broke completely loose and were blown out into space by the violent wind. This destruction, which proceeded with
infinite slowness, did little to impede the course of the
starship. The sail was constructed so it could function
even without the outer segments. So as the crew
watched on the viewscreen, they still believed there was
no direct danger. But they also knew they might lose
the hydrogen collectors any minute.

The starship, now slowly rotating about its own
axis, maintained its general heading while Captain Dolvar encouraged himself, and the rest of the crew, by
pointing out that as long as their course remained constant and the sails weren't totally destroyed, they could

still get into an orbit around D 40–31, another light year away.

Their second pass through the shock wave, five days later, came with surprising violence. Once through, they expected to leave the chaos behind, but their passage was far from smooth. Within hours of hitting the shock front an enormous section of the main sail slowly detached itself and drifted away, twisting and curling in the interstellar wind.

No one had imagined that the ship would find itself sweeping through a region of space containing so much matter. Soon the swirling eddies became too much for the remaining sail, which flapped and billowed in the great wind. Through the monitor camera the crew watched helplessly as the sail was slowly but inexorably torn to shreds. Three days later it came loose, together with the collectors, and drifted off into space.

Now the *Good Hope* was completely dependent on its inertia, the gravity of surrounding stars, and its limited hydrogen reserves. They were still moving surprisingly close to their original trajectory. During the separation from the sail, the entire ship had been mobilized for emergency procedures. Discipline had been maintained, except among the children, and all the life-support systems and production facilities remained intact.

Eleven months later, the *Good Hope* swung around to fire its retrorockets on the approach to D 40–31. All 237 crew members cheered as the good news was radioed off through space to the *Gaia*. All indications were positive and the new planet had been dubbed "Terra-2" by consensus, sight unseen. Close for the first time to the promise of ground beneath their feet, the *Good Hope's* star children had a wave of nostalgia for the home planet they had never known.

Without the sail their use of fuel had been woefully inefficient. Even if they achieved orbit around the target planet, they would have very little fuel for the shuttles needed to transport people and materials to the surface. There would be no chance of leaving for another system; the new planet's hospitality was assumed on blind faith and limited data.

Marret spent her days searching, computing, and analyzing. Dolvar devoted considerable energies to plans for the upcoming colonization. Nadra and Kara dreamed of children and pioneering. Barek and his crew checked and rechecked the ship's supplies and equipment. Everyone felt that luck was with them and the only remaining problems were technical.

The *Gaia* knew nothing of any problems.

On the third planet of the star D 40–31 the wind swept across the great plains. In the distance milky white clouds raced across the reddish sky. A giant creature with wide ears and a long snout moved silently on the dry surface, rooting for sparse clumps of grass. It hardly noticed as a bright fireball painted a trail of smoke across the sky. Its back was turned when, an instant later, a flash filled the sky and cast shadows of its form upon the ground. It bolted in panic when the sound of the explosion struck it with a deafening roar. In the distance millions of pieces of metal rained down on the plains, carving small craters. Within days the first layers of the desert sands, blown by the perpetual wind, began to fill in the craters and cover the fragments of alien metal.

SUPERNOVAE AND LIFE

CHAPTER FOUR

We are children of the stars. This is true in a very literal sense. It was in distant and ancient stars that the substances that make up all living things as well as the planets in the solar system were built. It took hundreds of millions of supernovae and planetary nebulae (another form of stardeath) to feed enough matter into space to arrive at the point where earthlike planets could form. This all happened during the 15 billion years or so since the universe started its wild dance outward. Plenty of time indeed. It wasn't until 10 billion years had elapsed that the sun and earth finally formed. Only after these events had played themselves out in the universal environment were conditions even remotely likely to lead to the emergence of life on our planet. Life on any other planet would develop in a similar way.

The events in "The Good Hope" obviously lie far in the future, and you may question whether it would ever be possible for us to travel to distant star systems and colonize habitable planets. Could we actually use star wind to travel from one star to another? Are binaries a threat to space travel? Above all, are there really other planets out there as hospitable to life as our own planet earth?

Consider the matter of propulsion by star wind. Nearly all stars blow out material in what are called stellar winds.

Seventeen Million Light Years from Home. Ink drawing.

These winds, which arise near the surface of the stars, get their energy from the interior. Some of the material in the top layers of a star literally gets blown out into space. Our sun produces a wind that blows at 1000 kilometers per second, but it doesn't blow out as much material as some larger and hotter stars. Some stars blow away so much matter that they can lose, in only a million years, as much mass as the sun contains. Clearly, if the sun were to do that it would be long gone!

The fastest winds observed blow several thousand kilometers per second, and although it is unlikely that our scenario of sailing in this wind could be realized, it would in theory be possible. However, even with an enormous sail, it would be hard to achieve a velocity greater than one percent of the speed of light.

To be practical, interstellar travel would require speeds that were at least a sizable fraction of the speed of light. Star wind, time warps, hyperspace—these are only a few of the many notions suggested by writers of science fiction. So far, all are in the realm of speculation.

Surprisingly, we know more, about the evolution of planets and how they might develop life. Astronomers calculate, assuming that our solar system is typical, that there may be billions of planets in our galaxy alone that are capable of supporting life. To find out how this can be, we need to look again at events in the first hours of the universe, which set the stage for the birth and death of stars and, most important for us, for the role of supernovae in the ultimate development of life.

Surpernovae, those cataclysmic explosions of stars, can be dangerous to living creatures in their immediate neighborhood. But they can also trigger the birth of stars and are essential to our emergence on this planet, indeed essential to the continued existence of life everywhere. How is this possible? The explanation concerns not just the explosion but what happens for thousands of years before and after the cataclysm.

The dust cloud known as the Coalsack, visible in the southern skies.
Consisting of enormous wreaths of matter that obscure the light
from more distant stars, this dust cloud contains essential material
from which future generations of stars will be born. The presence of
millions of stars in this photo also illustrates the staggering number
of stars that can be seen in certain directions in the Milky Way.
Photo by Ellis Miller, Cerro Tololo Inter-American Observatory.

The heavier elements, essential for making planets and living things, were not created in the first moments of the universe. They required the birth and death of stars for their creation. More important, the violent destruction of stars was needed to inject these elements into space and make them ready for use when future stars and planets were born. If this had not happened, they would still be locked up inside the stars, and we, along with our planet earth, would not be here now. The story of starbirth and stardeath is therefore basic to our life and all life in the universe.

You will recall that, in the clouds of swirling matter that spawn stars, two forces struggle against each other. The force of gravity pulls each cloud together, but the expansive force of the cloud's interior heat forces it apart. Gravity usually wins, and stars are born. But what triggers the starbirth? It appears that supernovae might well be the main character even at this stage in the formation of planets.

We can imagine that once upon a very long time ago an enormous cloud of gas and dust drifted between the stars, as so many other clouds of gas and dust do even today. That cloud collapsed and grew smaller and smaller until, rather suddenly by cosmic time standards, it formed a group of stars, each one accompanied by a flock of planets. One of those stars was our sun. But what actually made the original cloud collapse so suddenly? Let us go back through time and watch.

Space is filled with enormous interstellar gas clouds. They contain not only hydrogen, the basic gas that makes up most of the matter in the universe, but many other gases as well. Some are made of simple atoms such as sodium, calcium, or helium. Others contain molecules such as carbon monoxide, water, or ammonia. And pervading these clouds are solid particles, tiny indeed, that we call dust. You can see some of the dark dust clouds overhead in the bright Milky Way in the summer skies. The matter in the clouds is swirling about on itself. The molecules and atoms fly by one another at speeds of many kilometers per second, but the clouds are so many light years across that it takes millions of years for any given gas particle to traverse the cloud. When a particle tries to escape the pull of the cloud, it is drawn back in. The result is that the whole entity maintains a loose identity as a distinct cloud.

Gravity continues to pull the matter inexorably in on itself, so that the cloud of gas shrinks as time passes. However, another force opposes the pull of gravity. Collisions between particles inside the cloud build up pressure that acts outward, in the form of heat.

Some of this heat can radiate out into the cold surrounding space so the cloud can cool further. Then gravity, in principle, can win out over the expansive pressure and the cloud grows smaller. As long as the cloud can radiate more heat than it generates by its contraction, it will continue to shrink.

The force of gravity that acts inside this kind of cloud gets stronger as the cloud contracts, but just because the cloud is shrinking, the particles in it are closer together. Their random motion gets wilder and wilder, and the temperature of the cloud rises.

In nature it appears rather unlikely for the cloud to start contraction of its own accord, because its internal temperature provides enough energy to hold the cloud apart. Astronomers were forced to assume that somehow enough energy radiated away into space to allow the initial contraction. But now it appears that there may be another way to explain how such a cloud could start to contract and ultimately form a solar system.

Again imagine our cloud, light years in size, minding its own business in space. If now a nearby star explodes violently as a supernova, its explosion would send out a blast wave on a scale beyond anything we can comprehend. When that blast wave hits the cloud, it will pile up matter ahead of it as it expands outward. Thus the material in the cloud will suddenly find itself compressed. In that case the inward pull of gravity will become much greater than it originally was. Also some of the remains of the exploded star will mix in with this gas and give it a chemical composition that is different from that of the original interstellar gas cloud.

The nebula known as M 16. In this region of space stars have formed recently (several million years ago) and have heated the surrounding matter, left over after star formation, to incandescence. Dust clouds, within which new stars may even now be forming, are spread about the nebula. Individual pinpoints of light are stars mostly located between the nebula and us. Hale Observatories photo.

Because of the compression, gravity now can work very effectively in pulling the original cloud in on itself. The cloud shrinks quickly, and the pull of gravity gets even stronger. The cloud grows constantly smaller and smaller and hotter and hotter until, at some critical time, it breaks into many pieces that contract separately. These will become individual stars.

Because there was some movement of large masses within the cloud to start with, we find that the individual pieces of cloud slowly rotate, and as they grow smaller they spin faster. Individually, they are larger than our present solar system. The spinning cloud now flattens out (as any spinning object would), and we find that we have a hot, disk-shaped mass of material that is growing still hotter even as it grows smaller and spins faster.

The scene is now set for the birth of our sun at the center of this spinning disk, or solar nebula. It is likely that many other stars formed at the same time as the sun, but since the supernova was pushing things around pretty violently, all our near neighbor stars have long since wandered far off into the depths of the galaxy.

Only about half the stars we know (including those we see) are single. The rest occur as twins or triplets.

In most pairs, the two stars are so close to each other that seen from the earth they appear as a *spectroscopic binary,* which means that they look like a single star. The only way we can tell they are actually two stars is to study their spectra, or light patterns. From these, we can determine that there are two stars in orbit about one another. It takes measurements spaced over several years to make sure that is the case, but the spectra tell not only about the makeup of the stars but also what matter they contain and how hot they are. These spectra also tell how the stars are moving and permit us to infer their distance from each other. There are also binary stars that eclipse one another, as seen from earth, which means that first one star, then the other comes in front of its companion, causing an eclipse.

As far as space travel is concerned, the important and

somewhat frightening fact about binaries is that it is possible for material from the larger star to "leak" occasionally onto the other star. You can well imagine that such an event is not something to be taken lightly if you are in its vicinity. This leakage can lead to catastrophic consequences, because the in-falling matter heats up to enormous temperatures as it gets closer to the other star. This matter can even explode on its inward journey and be hurled out into space again, this time at great speeds. Often the bright flash produced by such an explosion is visible on earth. We call it a *nova*.

Some binaries are spaced so far apart that, in looking through earth telescopes, you can actually see the two stars physically separated in space. In these visual binaries, which usually orbit each other only once every few hundred years, there is no chance that matter would spill from one star to the other.

But let's go back now to the formation of planets. Modern-day observations by astronomers and earth scientists reveal that the earth contains more of the heavy elements than are usually found in stars and interstellar matter, an observation that can best be explained if we picture a supernova as the trigger that led to the collapse of the solar nebula gas cloud. The heavy elements made in the supernova explosion polluted the gas that was originally there; hence the sun and earth contain the same chemical elements.

The first generations of stars had only hydrogen and helium in them, and there was none of the matter needed to form solid planets, nor were there supernovae to trigger their birth. The universe originally did not have elements like iron, silicon, magnesium, nitrogen, and oxygen that we now find all about us. Thus the early stars could not have had earthlike planets. It was only much later that the heavier elements were formed. This happens when the heat in the core of a star grows so great that helium fuses with helium to form carbon, and the carbon then fuses with carbon to form oxygen. As the star grows even hotter, heavier and heavier elements can be created in its interior. And so the basic building blocks necessary for a planet's formation are made inside distant stars. Even then, only stars that explode as supernovae can release these elements and send them out into space.

There they mix with other stellar debris and are available for further star formation. Only after millions of supernovae have occurred in our Milky Way galaxy will there be enough of the heavy elements to form planets like the earth, and life as we know it. We are all fossils of stardeath. Each atom in our bodies was formed in a star, somewhere, sometime.

Not only the atoms of our world but also the complex molecules of living organisms depend at least partly on supernovae. In Chapter 2 we mentioned the bad effects that cosmic rays could have on us, but like many things in nature they can have beneficial effects as well.

In the twentieth century it has been discovered that energy or radiation can take many forms. These include radio waves, heat, light, ultraviolet, x-rays, and gamma rays, all of which are found throughout the universe. They differ from one another only in wavelength. Radio waves have a long wavelength, centimeters to hundreds of meters long, while light has a short wavelength around a hundred-millionth of a centimeter. X-rays and gamma rays are shorter still.

Until about a hundred years ago the only form of radiation recognized was light. Everyone was aware of heat as well, but no one realized that heat was closely similar to light, differing only in its wavelength. No one then knew of the existence of radio waves, x-rays, and gamma rays. The point is that until this century we would have been at a loss to describe those initial seconds of the universe when all was radiation, since the radiations involved were not part of our vocabulary. The closest description that ancient peoples could achieve, given their limited scientific knowledge, would have been for them to say, of the early seconds of the universe, "In the beginning all was light!"

From the instant of the big bang, radiation traveled outward at the speed of all radiations, which is the speed of light. Galaxies formed, within which stars were born and died. Some died by exploding violently as supernovae. These supernovae produced cosmic rays that subsequently may have had a profound influence on the emergence of life, in that they could have been the indirect cause of electricity in lightning storms.

Maybe this sounds far-fetched, but the connection runs somewhat as follows: First, we know that lightning flashes occur when large amounts of electricity build up in clouds. At some point the voltages become large enough that enormous sparks, the lightning flashes, jump from cloud to ground or from cloud to cloud. But where did that electricity come from? The answer has been sought for many years, and one recent picture involves the action of cosmic rays coming from the depths of space.

When a cosmic ray strikes and shatters a molecule of air, the molecule is broken into positively and negatively charged fragments. This gives rise to static electricity. On a clear day the air around us is known to be filled with such electricity. It is also known that the positive and negative charges separate out. The negative charges apparently favor the ground, and the positive charges move upward in the air. On a typical day there might be a 200 volt difference between your head and your toes. Since this is static electricity, we will never feel a shock from it; the energy in these 200 volts is too small.

A chemical effect, known as adsorption, leads the negative charges to stick to any surface with which they come in contact, while the positives don't stick very easily to anything, so they just float about in the air and are carried around by updrafts. In the presence of strong updrafts, such as those producing cumulus, or storm, clouds, the charged particles continually being created by cosmic rays are swept up into the clouds as soon as they are formed. The negative charges stick to the water droplets at the bottom of the cloud, while the positives are swept higher up in the cloud. This charge separation causes the buildup of enormous voltages in the cloud, leading to a lightning discharge when the voltages are too great. The interaction between the electricity in the

Primeval Landscape. Oil on canvas. The painting illustrates the concept that the planets of the solar system formed from the accretion of rocks and particles of all sizes that orbited the newly born sun.

bottom of the cloud and the ground also allows lightning to pass from cloud to ground.

Apart from the obvious fact that lightning kills people and animals, it also starts forest fires. Nowadays, foresters are becoming increasingly aware of the fact that naturally occurring forest fires are important for the continued health of a forest and the larger-scale ecology of our planet. In some national parks they are allowing naturally produced fires to burn themselves out, so as not to interfere with the ecology of the forest system. If cosmic rays do play a role in generating lightning, then we can even relate what has happened in distant stellar explosions over the last billions of years to the health of forests (and all life?) on our planet, since most of our oxygen comes from green plants. Life on earth is far from isolated from events out in space.

A further importance of lightning to life is that it may well have been one of the main sources of energy needed for the creation of life on the early earth. Lightning discharges, together with ultraviolet from the sun, provided the necessary energy for chemical processes to take place, processes that ultimately led to the emergence of life. These chemical processes allow more and more complex molecules to be built up from simple atoms. Molecule building requires energy, and lightning is a good source of it. Cosmic rays may provide the electricity for lightning, and supernovae are a good source of cosmic rays. It is a remarkable chain of events.

Cosmic rays are produced during the explosion of a star, but they are also produced for thousands of years afterward. Why is this so? The answer lies in the rapidly spinning neutron star (sometimes called a pulsar). This is the remains of the central part of the star that collapsed to a solid ball of neutrons. The pulsar is an extremely regular ticking clock. We receive a pulse of energy from it every time some particular region on its surface is pointed at us. We can imagine the pulsar as a lighthouse beacon in space, its beam sweeping by at highly regular intervals. Some pulsars pulse once every few seconds, others many times a second. A pulsar continues to feed energy into its surrounding supernova shell as it spins.

It can do this because some of the energy of its own rapid rotation can be fed into the surrounding nebula through a mechanism that involves the magnetic field of the pulsar and the nebula itself.

Pulsars have enormously strong magnetic fields, reaching a billion times the field we have here on earth. As the pulsar spins; this field gets wound up like a watch spring. However, magnetic fields do not like to be wound up, and they tend to realign themselves by forming little loops of field that can separate themselves out from the rest of the field and go hurtling off into the surrounding nebula. As a side effect, the particles dragged along with the field also get injected into the nebula, thus maintaining a constant source of cosmic ray particles. These are later leaked out into space beyond the nebula and ultimately permeate all of space in the galaxy.

A pulsar cannot continue to throw off bits of magnetic field forever without losing all its own energy. Indeed, it will slowly run down as it loses this energy to the nebula and this slowing down has been measured by astronomers. In the best known pulsar, the one in the Crab nebula, the loss of rotational energy just balances the energy being emitted by the nebula itself, suggesting that the pulsar is the source of all the energy radiated by the nebula. This explains why the nebula has continued to shine for the last 900 years, ever since it was seen to explode in 1054, and why it will probably continue to go on shining and expanding for thousands of years to come. Needless to say, other stars will be engulfed by this expanding shell, with serious consequences for any life on their planets.

In summary, supernovae are the source of heavy elements needed for planetary formation and of cosmic rays apparently needed for the evolution of life. But what if the evolution of stars had been different? What if the heavy elements that were formed were never fed back into space by stellar explosions? One thing is clear; there would now be very little new star formation taking place because there would have been no recycling of matter back into space. Any star that did form would be made only of hydrogen and helium, and none would have solid, earthlike planets in orbit about them. These stars would ultimately die their quiet deaths, as many do even now. They would become cold hulks with equally cold gaseous planets orbiting about them. Life would never have been given a chance, since the elements needed for life would never have been recycled into space. Fortunately for us, supernovae did happen, and we are here now.

MAGNETIC FIELD ZERO

CHAPTER FIVE

Some of the disastrous effects of a supernova light years away from earth could also occur as a result of events *inside* our planet. To understand how this is possible, we must look at the earth's magnetic field, which appears to play an important role in sustaining life. It shields us directly from the constant wind of particles blowing out from the sun and from most of the cosmic ray particles from space.

The earth acts like a giant magnet whose effects can be felt a long way out from our planet. The region of space around a magnet in which such effects can be felt is called its *magnetic field.* Only recently has the precise nature of the earth's magnetic field been probed. Near the earth, the field is shaped much like the field produced by an everyday bar magnet. However, the interaction between the earth's field and the solar wind modifies the field far out in space above the earth. It is this that makes the study of the earth's field so interesting, but to understand it we must go back another step to see why the earth has a magnetic field at all.

The existence of the earth's magnetism appears to be linked with the presence of a molten core in the center of the planet. Due to a chance set of circumstances there are motions within this molten mass that set up little whirlpools called *eddies.* Because these eddies contain electrons they can generate their own small magnetic fields, much as electrons flowing in a coil of wire do.

The fields from all the eddies add up to generate the total field of the earth. At present the eddies flow so that the earth's field is directed primarily in the north-south direction. The magnetic field could equally well have been oriented opposite to what it now is, and it is now known that it *has* been the other way around many times in the course of the planet's history. The earth's magnetic field has flopped first one way and then the other at least 20 times during the last four million years! If there were as many eddies rotating clockwise as counterclockwise, the magnetism produced by these eddies would cancel out and the earth's field would be zero. This situation seems to occur between field reversals and it is believed that the field may remain zero for 1000 to 10,000 years during such times.

These reversals have been detected by studying particular sedimentary rocks deposited layer upon layer by the sea. The ages of the layers are dated by examining the amount of certain radioactive materials in the rock and by knowing how much of a particular element is produced by radioactive decay in a certain time period.

Measurement of the direction of the magnetization of rocks in sedimentary layers (as well as in lava flows that occurred sporadically in the past) reveals the magnetic field reversals. Magnetism can be "remembered" by rocks formed under various conditions. In the case of molten lava flowing in the presence of a magnetic field, the molecules and crystals in the rock can be aligned a little by the field. When the lava solidifies, the parts of the rock freeze into the preferred orientation. From then on, the rocks have their own magnetic field, a record of the field present when they solidified. Lava flows show that at different ages in the past the earth's field was not always pointed in the same direction as it is now.

While we know that the field has reversed many times, present-day theories are nowhere near explaining or predicting such reversals. The only thing we know is that the main field is currently declining, and, taken at face value, may start to reverse itself 50 to 150 years from now. Since detailed measurements of the earth's field have only been made in the last two decades, no one is sure what the changes in the measured field really mean.

Viewed from space, the earth's field is seen to interact with the solar wind, thereby changing its structure considerably, especially as it gets farther from the earth and violent events on the sun send clouds of particles past the earth.

Back in the late 1950s, several satellites were sent into earth orbit, and one of these carried a counter designed to measure the number of cosmic rays reaching us from outer space. In earth orbit, the Explorer cosmic ray detectors started to count enormous numbers of particles, and then the counters failed, or so it appeared at first. Later the counters started working again, and again they counted very large numbers of particles. The count rates seemed to depend on where the satellite was in its earth orbit. Then it was realized that the counters hadn't failed at all. They had simply been bombarded by too many particles to count, failing only when the levels were too high.

Subsequent experimentation revealed that the earth is surrounded by two doughnut-shaped regions in space in which large numbers of cosmic ray particles are trapped. These regions of high particle density, named after their discoverer, James van Allen, are called the *van Allen radiation belts.*

The origin of the particles in the belts was once believed to be the sun. Now it is thought that the particles in the lower belt (located about 3000 kilometers above us) come mostly from the earth's atmosphere. High-energy cosmic rays from space ignore the earth's magnetic field and hit the upper atmosphere, where they shatter atoms and molecules, producing charged particles that will then be drawn to the magnetic field. A considerable number of these particles filter up into the van Allen belts. The upper belt (located 15,000 to 20,000 kilometers out in space) contains charged particles that originated in the solar wind and then leaked into the magnetic field of the earth.

We can think of the earth as a ship moving through a sea that is the solar wind. Just as a moving ship sets up a bow wave ahead of it and leaves a wake of swirling water behind it, so it is with the earth and its magnetic field in the solar wind. The wind is constantly blowing past us, as if we were moving into a stationary cloud of matter tied to the sun. As a result, a bow wave is set up ahead of the earth, where the solar wind first hits its magnetic field. This occurs in a region called the *bow shock,* located between the earth and the sun, about 80,000 kilometers out from the surface of the earth, where the solar wind is deflected and forced to swirl past the earth. The magnetic field of the earth acts as a shield to the solar wind particles, preventing the particles from colliding with our atmosphere.

As the wind gets deflected around the earth, it streams past and takes on a pattern something like the wake of a moving ship. This produces the earth's "magnetic tail." Events in this tail are of considerable interest to us on earth, because particles from the solar wind can now leak into the weaker magnetic tail and run along the field lines until they crash down toward the earth's poles. When these particles hit the atmosphere they heat up the gases there and light them to incandescence to produce the beautiful phenomena we know as aurorae.

It is important that the magnetic field of the earth protects us from the solar wind and the clouds of particles that are sent out by explosive events on the sun. Without this magnetic field, the solar wind and particle clouds would strike the atmosphere directly. It has already been shown that such particle collisions with the atmosphere are able to produce chemical changes that lead to a reduction in the ozone content. Since the ozone layer is our only protection from the ultraviolet of the sun, we would be exposed to the same hazards associated with particle blasts after a nearby supernova explosion—sunburn, skin cancer, and even death. The earth's magnetic field is in fact shielding the ozone layer from damaging particles thrown out by our own sun.

We can only speculate about what will happen to life on

this planet the next time the field drops to zero, en route to a total reversal. The last time this happened is believed to have been 600,000 years ago, so we have no direct experience of an event of this kind, although the extinction of a few simple creatures has now been related to such field reversals.

We must assume that evolution struggled on despite nearby supernovae *and* despite magnetic field reversals. Yet earth's fossil history shows the disappearance of whole families of living things within relatively short times. Over the last few decades scientists have put forward several theories to explain the disappearance of species, but never to everyone's satisfaction. Now a theory has been proposed that links the known reversals of the earth's magnetic field to the disappearance of several species of simple life forms. The absence of the earth's magnetic field for periods of up to 10,000 years may have hampered life seriously.

It has been convincingly shown that the extinction of several simple species of fauna (in particular the order of Radiolaria, single-celled marine organisms) happened at just about the same time as magnetic field reversals. Ths disappearance of animal life is revealed by studying the fossil record (like the field polarity record) in various rock layers and noticing that at some time in the past some type of fossil was present, whereas several thousand years later the fossils did not occur in the rocks. As far as we know, living things are not directly sensitive to magnetic fields and certainly not to the extent that the mere presence or absence of fields kills them. There had to have been some secondary cause. Evidence is accumulating that the important events took place during times of field reversal, when the magnetic field dropped to zero. At those times, the unshielded earth was susceptible to the chemical destruction of its ozone layer, as described in Chapter 2.

There is yet another phenomenon that could act to remove the ozone shield during field reversals. Solar flares, explosions on the sun's surface, produce very large numbers of particles, especially protons. The magnetic field of the earth usually deflects these solar protons, but some of them get

Devonian Landscape. Oil on canvas. Life forms of the Devonian era, 400 million years ago, populate this surreal view of the early earth.

trapped in the van Allen belts. If the earth's magnetic field were removed for a few thousand years, we should expect that a violent explosion on the sun would send protons crashing down over the whole earth rather than just near the poles, where they would otherwise be guided past by the protective field.

The earth apparently becomes very vulnerable to proton storms originating on the sun during the times when our defenses, usually provided by the earth's magnetic field, are down. It is during such times in the past that primitive organisms such as the Radiolaria have been wiped out. Calculations have now been done which show that this mechanism can work. The effects of the protons in producing electrons and hence nitrogen oxides and a reduction in the ozone layer can be well simulated in computers. This chemical destruction of the ozone layer was one of the chief objections to large numbers of SSTs flying all over our planet, since these planes produce the same harmful nitrogen oxides in the high atmosphere.

The susceptibility of simple organisms to destruction by ultraviolet light has been well measured in the laboratory, and it is found that many simple organisms exist at levels where just a small increase of ultraviolet can kill them. If these organisms were killed off, the consequences to our ecological system, where one organism depends on so many others for its survival, could be far-reaching. As yet no one knows just how far-reaching the effects might be, but work is proceeding in various laboratories to find out.

We might now consider the case of the disappearance of the dinosaurs. Did that occur as a result of a field reversal that laid the planet bare to particularly violent solar explosions? No one is sure, because detailed measurements of the earth's magnetic field unfortunately do not extend far enough into the past. However, many species must have been affected by events on the sun during earth's history. Besides, in the past, the sun may have been much more active. This is a tempting thought, for the dinosaurs as a species lasted for approximately 200 million years and must have survived many field reversals without incident during that time. It would be difficult to explain why they went extinct only 70 million years ago, even if the time of their disappearance was correlated with a field reversal.

What about the next field reversal? No one knows when it might start. At present the earth's magnetic field is changing very slowly; if the decrease observed now continues at the same rate, the field could drop to zero in a few hundred years or less. We would then once again be particularly vulnerable to proton storms on the sun. Whether or not a species such as *Homo sapiens* would suffer is unknown, except that we can take solace from the fact that we are a very adaptable life form, quite skillful at controlling our environment, and dependent on a variety of food supplies.

The sensitivity of the ozone layer, so essential to our survival on earth, is only just being recognized. Many "pollutants" can remove the ozone, even when the magnetic field is stable. In the early 1980s a satellite especially designed to study the ozone layer will be launched. It should tell us a great deal about how the ozone layer is even now affected by the sun and by pollutants introduced by civilization. That should at least enable us to work toward preventing accidental destruction of the ozone layer The most interesting thing to be discovered by the new satellite is how the sun affects the ozone layer even in the presence of the earth's protective magentic field. At times of sunspot maximum, which happens every 11 years (the next maximum being around 1980–1981), there will surely be many solar flares that might shoot clouds of particles streaking toward the earth, thus sending the ozone layer into paroxysms.

If the terrestrial magnetic field should simply disappear, then the first major solar flare thereafter that sent out a lethal cloud to earth would start the process of ozone destruction. The ultraviolet doses received on the surface of the earth would go up, and the incidence of skin cancer would no doubt increase. If there were a series of massive solar flares, then the ozone layer might be mostly removed, a catastrophe for all of us. Exposure to sunlight would become fatal. There would be no more sunbathing, walking about unprotected from sunlight, or skiing without ultraviolet filters over all exposed parts of the body. Many organisms exposed to ultraviolet radiation would quickly die, and we would probably enter an era where virus mutations would far outstrip our ability to cope with them.

If the magnetic field reversal data are taken at their face value, there appear to have been long periods of about a million years over which the field stayed substantially the same. But there were short times during that million years when it flipped back and forth about every 200,000 years. However, the past 600,000 years seem to have had no reversals at all. We might speculate that while regular field reversals would encourage the survival of species that could best cope with it, a period of 600,000 years without a reversal might precipitate larger evolutionary changes when it finally does occur. This is because life is no longer adapted to dealing with an increased proton dose or to increased ultraviolet radiation.

Earth's magnetic field has certainly facilitated the emergence of life and its survival. This field is our protection from the constant blast of particles emerging from the sun, a blast that is punctuated by storm clouds driving into the earth's magnetic field boundary. Traces of these storms on the sun manifest themselves as beautiful aurorae here on earth, as we have seen, but remove the field and we can be certain that the course of life will once again follow new paths.

ICE AGE
CHAPTER SIX

One shovelful after another, Ben Tibbets slowly cleared the snow off his walk on the Caribbean island of St. Martin. He resented having to deal with the stuff in such a primitive way. A hundred years ago his great-great-grandfather would have had a functioning snowblower, but his had been rusting in the garage for five years now. You couldn't get parts or replacements for anything these days.

The street was clear enough. There were still enough skimmers riding their warm air pads to melt the snow. His own skimmer was wearing out, but he could still use it for emergencies. The sounds of construction equipment drifted up the street from where the harbor had been 10 years ago. Now they were building an enclosed residential mall there because individual homes were just too hard on energy supplies; the new prefabricated harbor had been built farther out on the receding coast. When the mall was finished, they'd have to give up the comparative luxury of their own house. He and Mara would be confined to a convertible living room, and all the kids would be sharing a single bed-room, assuming the kids were still here. He threw the last few shovelfuls with a vengeance, remembering when it wasn't so critical to preserve body heat and Tom could shovel with him.

Putting the shovel down at last, Ben stepped quickly through the thick door of his foam house, thinking how it looked like a giant igloo with all the snow on top, and began stripping off his thermal cov-eralls.

"Daddy, daddy, there's a show on about New York City! They're talking about what it was like when grampa left."

"Not grampa, Jenny, *great*-grampa. You go watch it and I'll be there in a minute."

He held the spigot of the kitchen sink in his frozen hands as it heated and then carefully let just enough cold water through it to make a single cup of chicory. When he got to the living room, Jenny, Tom, and the baby were all assembled in front of the wall screen, which showed an aerial view of the Brooklyn Bridge, deserted and ice-coated. He recognized its unique struc-ture from pictures. The camera moved on to the tips of huge old buildings protruding from rivers of ice and snow. The narrator was talking about the 300-foot drop in sea level since the twentieth century as the video crew's helicopter swung over the hulks of ships show-ing through the ice in the former harbor, where sweep-ing winds kept the ice low enough to show the wreck-age.

"Daddy, tell us about great-grampa again. Did he really live there?"

"He sure did, but let's watch the rest of this and then I'll tell you. Look, they're showing pictures on the ground now."

The documentary team was now snowshoeing and skiing in sub-zero winds as they battled their way over the former streets of Manhattan, and the narrator continued with some historical facts. "The winter of 1976–1977 was a foretaste. A few years later our grandparents remembered 1977 with nostalgia, not noticing that the summers were cooler, too. The fuel shortage reached critical proportions as winter after winter stayed cold. The sunspot cycle was combining with the onset of world cooling to produce very cold winters, even though the average temperatures were dropping by only a fraction of a degree a year. But the next ice age was on its way. The glaciation had started.

"Cold and the scarcity of fuel weren't the only problems. The period of unusual warmth that lasted from the 1950s through 1975 was accompanied by enormous population growth that was barely matched by food production. But scientists expected the warm spell to end soon, since it was apparently related to the sunspot cycle. By 1988 it was evident that society could no longer cope with the change in weather combined with the energy crisis, nor with the massive famine that resulted. The warm spell was over."

Now the documentary was showing old footage from the early twenty-first century of rioters and looting sprees. Rubbish had piled high in the streets as removal systems broke down, and people abandoned their homes for warmer climates. A short sequence went back about 100 years from the present—to the beginning of the real unemployment crisis and the massive flu epidemic of 1988—and then the show was over.

Ben turned to Tom and Jenny, who were still waiting for his story of their great-grandfather. Mara had come in during the show and was holding the restless little Acadia on her lap. She had found the baby's

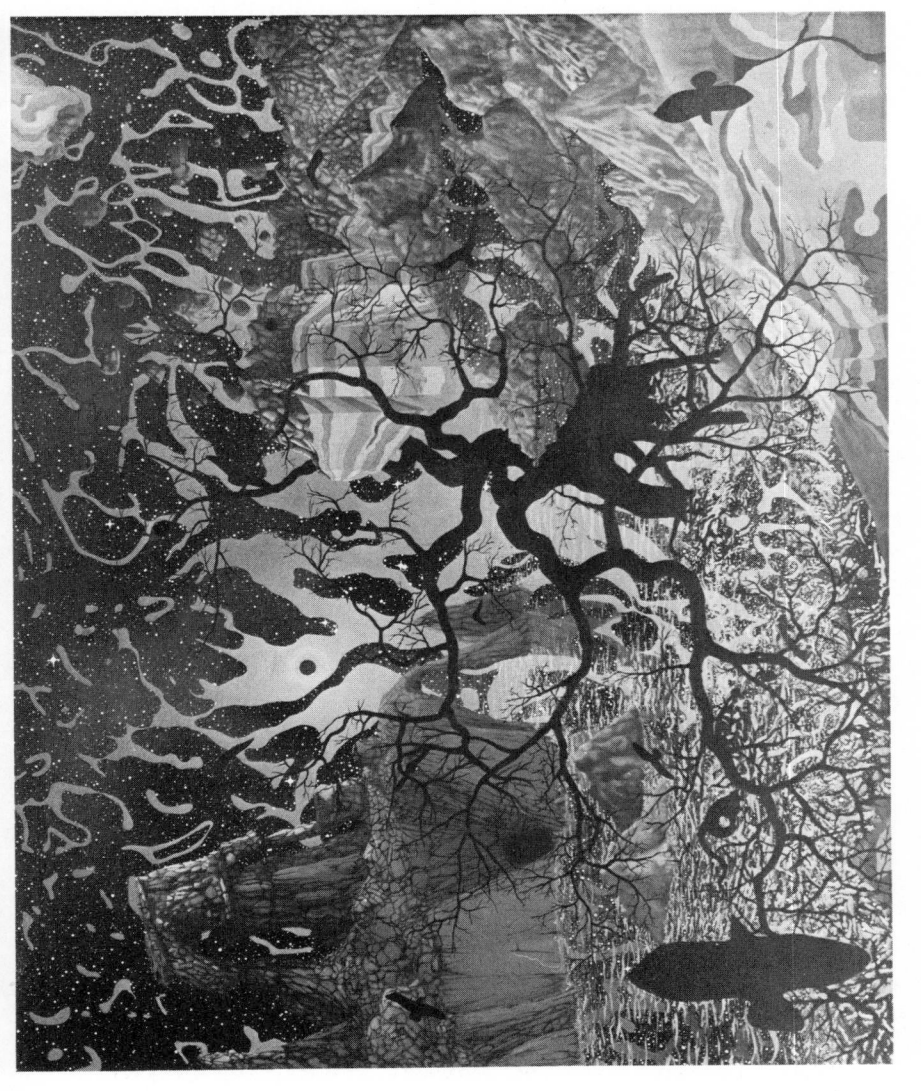

Rosetta Cove. Oil on canvas.

American Indian name in an old book; it meant *abundance* and they had given it to her in a spirit of hope against hope. They weren't supposed to have more than *one* child, but Ben had pulled strings at the Population Department in San Juan, now the U.S. capital.

The older children were tugging at him, and he began.

"Right where the show ended is the time your great-grampa left New York City—in 1988. He was 18 then, and there was just no work. The cold weather had started more than 10 years earlier, and the energy crisis got worse all through his childhood. They didn't know what a crisis *was!* Then the flu epidemic. . . ."

"Just get to the part where he left, Dad. We don't want to know all that plocky," Tom interrupted.

"Well, hold on, I'm getting there. His father refused to believe the warm weather wouldn't come back, even when the first evidence came out that the northern glaciers were starting to move. A lot of other people had already gone south, and my grandfather decided to leave home on his own when the flu epidemic killed his uncle and two of his cousins in 1988. He was just 18 and didn't have any money, but he hitchhiked to Florida and managed to find a job.

"He used to tell me how the northern companies were all going bankrupt and how things got bad even in Florida as the weather got colder and colder. Pretty soon everyone believed what the scientists were saying, that we were in for a long ice age that could last tens of thousands of years. The government moved from Washington to Puerto Rico in 2035, 45 years ago—but you should know that from school. My grandfather was 65 then, and my father—let's see, he must have been about 36, but I wasn't born yet. Anyway, my dad decided to follow the government south and came here to St. Martin, where you and I have lived all our lives." He smiled at them.

"What's going to happen to us, Daddy?" Jenny asked anxiously. "Is there anywhere left to go?"

"There are some places, Jenny, but most people aren't allowed to move anymore, only the luckiest ones." Too soon to say anything more, Ben told himself. "It's getting harder and harder to get enough energy out of the solar planets, even with the population down this far, but the government's doing the best it can. This morning's news said there were more greenhouse failures in Puerto Rico, so we probably won't see any more fresh vegetables."

"Are we going to starve?"

"No, sweetie, we'll manage somehow," Mara broke into the conversation. "We'll have to get used to lots of new changes, but we've done that already! The new apartment will be fun," she said, mustering cheer. "At least your friends will be where you can see them all the time, and you won't have to go outdoors to get to school."

"Mom, can I go out to that new wreck they found below St. Kitts? Dan and his father said they'd take me, and we might find something really valuable to sell and. . . ."

"No, Tom, you certainly can *not*," Ben said sternly, although he had clearly not been addressed. "Three more people were killed out there yesterday. The tides come in and out as much as 12 miles and they're too dangerous for wreckhunting. Besides, no one's finding any big treasure."

"But I'm 15, Dad. I can take care of myself."

"You can take care of all of us right now by setting the table while I put supper into the stove. The casserole, Mara?"

Later that evening when the children were all asleep, Ben and Mara had a serious discussion about their future. They felt lucky to have three children. The World Crisis Organization program for eventually reducing the earth's population to three million people spread around the equator, where there was still enough solar power to warm them, was obviously proceeding on schedule. And their little house was snug enough, with near-perfect foam insulation and five bodies radiating warmth. They scarcely needed power for heat, only for lights, hot water, and appliances, including the computer terminal for his work. But they couldn't stay here much longer. He pulled Mara down to him on the built-in foam deck with its comfortable old pillows. Cozy as it was in here, they often found themselves sitting very close now. The shallow-domed roof creaked overhead.

"You know, I think we may be getting out of here just in time. That load of snow and ice is going to break this old house one of these days."

"I know, Benjy"—she was calling him that more often now—"I don't want to move, but I know it's got to be. It's just that we'll be so crowded, and there's no future for the children."

Ben decided this was the time to air his new proposal. "Mara, I want to try and get the kids to Quito. That's sure to be one of the last outposts."

Mara pulled herself away from his shoulder, her long, still-black hair catching on the code-key chain around his neck. For an instant he caught a wide-eyed look of fear before she composed herself and smiled bravely. "If it's got to be," she said, repeating her earlier words, "then it's got to be." In fact, she had been thinking the same thing for a long time. "Tom and Jenny have grown up fast. If we can put them in good hands, I know they'll be all right, awful as it will be for us. I don't know what I'd do if they hadn't let us have Cady. At least we'll have one child here."

"I'm talking about Cady, too, Mara. There won't be another chance."

"Cady? Ben, she's not even two! You can't mean it, she's a baby!"

"I do mean it, dear. I've had the application for all three of them in for a year now, and the approval came through this morning on the terminal. I didn't want to tell you while the kids were around."

"But, but. . . ."

"We've just got to be realistic. Time is running out for us here, and this is the last group going down to the equator. They're only sending children now, with enough adults to care for them. In Quito they'll go to survival school and stand a good chance of getting into the Shelter Program. From what I've been able to gather, it looks like the suspended animation systems down there have been perfected. Almost 100 years for some of the. . . ."

"Oh, Ben, what are you saying?"

"I'm saying that that may be their only hope some day. We're not just waiting for the earth's orbital wobble to straighten out, we're in a cosmic dust cloud! It could be a thousand years or a hundred thousand years. No one wants to believe that, but it's true, and nothing can be done about it. We're all going to freeze, but only a few will ever stand a chance of waking up to a warm earth. I want our children to be among them."

"If they've really got that technology now, why can't we all survive?"

"Because we can't build enough of that kind of equipment here. We're lucky to have warm housing for our natural lives. Suspended animation won't be used until the bitter end, but I believe that end is nearer than we think.

Mara leaned back against Ben, where he couldn't see the tears running down her cheek. She wanted to rush to the children and hold them fast, but the best thing to do was just sit here and try to think this through. If Ben was right, how could she oppose the plan? She closed her eyes and saw Acadia's tiny form snuggled under the covers. Cady, with her own dark hair and Ben's smile. Then she saw Cady in some strange-looking device with computers attached. The image was of hospital equipment, with Cady looking pale, immobile, and tiny. But, no, she'd be grown up then. Mara tried to revise the vision, then pulled herself out of it. It was too awful.

"When? When will they have to go?"

"In about a month."

Cady rolled over in her bunk, opened one eye, then tucked her long hair back into her hood so she could read the chronotherm. It glowed ominously back at her: 5° C. Almost. What did she expect, she asked herself, that's what it said every morning. She reached instinctively for covers, as in her childhood, but there was nothing to pull up; her thermal suit did all that could be done. She might as well get up and do her instrucheck; at least she'd stir up her own heat by moving around.

Something registered as she sat up. Turning back to the chronotherm, she tugged at her memory: JAN 8 2143 6:25 5°. Jenny's birthday! Ten years older than herself, that would make Jenny 71! If she had managed to stay alive, that is. Cady hadn't heard from her or Tom for several years now. Was she the only one left? Her parents were certainly gone, as they never could have gotten permission for age-arrest. It was depressing not to know anything, not to know if you were the only one left in your whole family.

Cady pulled herself away from the dangerous thought. Depression had to be fought. There were things to do.

She got up and went straight down the corridor to the solarium, nodding to Buff as he passed in the other direction. Solarium, she thought cynically, Why did names like that persist? All the lights over the plant beds and animal cages were artificial, like all the other lights in shelter E-013-749. As she opened the door, she saw it was still dark. Her chrono had said 6:25, so it could only be another minute before the lights came on.

Groping her way to the computer, she heard pitiful squeaking. Doesn't sound good, she thought; something else is dying. The panel lights guided her to the corner, where she slipped into the seat and started calling up data just as the overhead lights flashed on the dawn. Cady winced. Why had she reacted that way all her life? Was there something innate in people that made them expect a gradual dawn?—and a nice warm environment? Why couldn't they give up these things and adapt? She punched viciously at the panel, telling herself she just had to cut out this kind of thinking. Maybe it was age. But that was unreasonable, for at 61 she was physiologically only half that age. The implanted monitors told her that every day, and the mirror confirmed their numbers.

It was in section 14, the gerbils. Three of them dead. Their monitors registered zero on all counts. The rest were probably fighting over the remains. She'd had more hope for them than this; the species had evolved on cold tundra, after all. Maybe she'd better suspend the rest now.

Most of the animals remaining were at the Big Shelter, but a few were spread out among the others. There was no way to tell any more which shelters might survive the crushing ice, bearing down relentlessly.

Communications with all but a few had broken down some time ago. Cady sighed and looked up at the old photo on the wall over the panel. Luxurious tropical growth steamed in the sunbeams that filtered through the dense foliage, and brightly colored birds swung on the branches. It was impossible to imagine, like some other planet. She had never seen a bird. Noah's ancient Ark obviously did better than they were doing!

When the data were all recorded and the surviving gerbils had been taken down for suspended animation, Cady hurried to the lounge for breakfast. Dropping a tablet into her cup of hot water and unwrapping her protein bar, she settled into a snug egg-chair. But something hard was wedged down into the padding. She reached for it and pulled it out. It was a handwritten log of some kind. She held it for a moment, wondering if it were private. But what was privacy any more? Besides none of the others had finished their morning routines yet, and she was alone for the time being.

Keita's handwriting gave her some trouble at first. It was neat enough, but she hadn't seen any for a long time, and never very much of it. She was surprised that Keita had bothered learning, but as he said on the first page, "Technology might be crushed beyond repair one day while a little book survives in some crevice." Cady smiled to herself, recalling the many times Keita had expressed this view. Like herself, he had been given an ancient name that sprang from a pre-technological civilization. The original Keita had been an African hero who built an empire out of a small state before Europe reached the Middle Ages. The Keita here at the shelter would be one of the last survivors on the downhill side of civilization. Ironic. Cady stopped musing and read.

The first part described the shelter. She leafed through it quickly as it was only a catalog of equipment

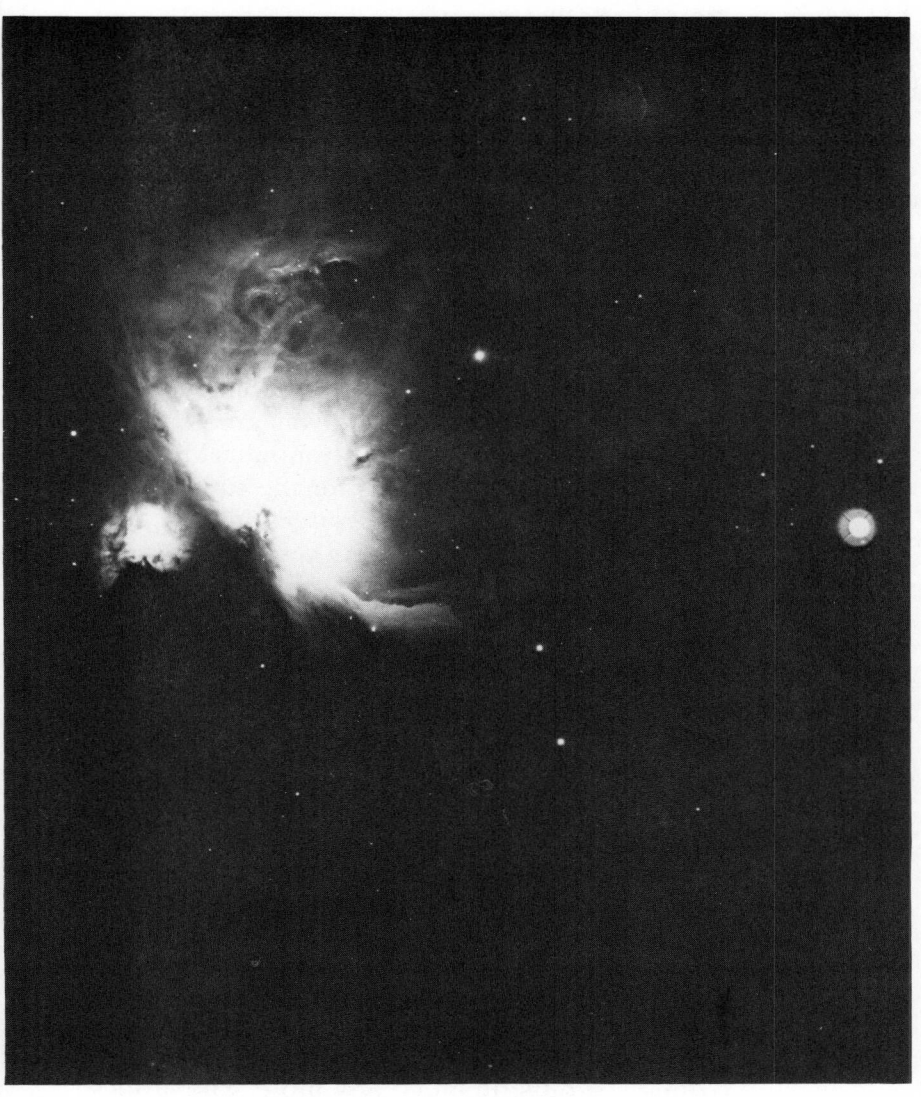

The Orion nebula, a cloud of hot gas surrounding four bright stars known as the Trapezium (not visible in this photo), located inside the brightest part of the nebula. Swirling dust clouds are visible all around the nebula. This photo, made with the new 150-inch telescope at Kitt Peak, shows very few stars because of the short time exposure. Photo Kitt Peak National Observatory, operated by Association of Universities for Research in Astronomy.

and procedures. Then there was a summary of the ice age onslaught:

"Around 1990, a small observatory in the midwestern United States concluded that the sun was entering a galactic dust cloud. Spectral analysis of nearby stars showed the characteristic signature of interstellar dust in large quantities. Up to that time, it had been assumed that the colder weather was due to orbital irregularity, and the new information was greeted with skepticism even by the American Astronomical Society. When confirmation came, estimates for the severity and length of the coming ice age were revised upward. For the first time, many scientists became convinced that the survival of earth life was seriously threatened.

"Studies showed that the gradual temperature drop actually began in the 1940s, although the '60s and '70s showed a deceptively warm spell. The quality of astronomical instruments before 1990 was too poor to observe the cloud to which earth had been responding for decades. Whether the main effect in the reduced sunlight at this time was due to orbital variation or the cloud was still uncertain, but the cloud was sure to have a longer-range effect. It was then visible against the summer triangle of the brightest stars.

"Now the night sky, what little we see of it, glows with the reflected light of the sun's rays. But the sun and its nine helpless planets are still plunging deep into this dense globule of dust that controls our lives. . . .

"We don't get outside much any more. The airlock is too costly to use more than once every few years, and then only for necessary checks. But there's nothing out there but snow and ice and harsh winds. The sun still shines, of course. It's almost impossible to believe the year-round average temperature has only fallen nine degrees from the time when jungles flourished here!

"A recent communication from F-104-629 leads us

to believe they are near the end, although they tried to sound optimistic. Only the equatorial stations like ours have any real hope, and I'm getting reluctant to say even that. The others can't fight the glaciers any longer. All their resources are going into maintenance of the life-support systems. Consolidation of some of the smaller shelters has made the work load more economical; the Big Shelter has thousands of people now, compared to our 322. They say they can survive another 3000 or 4000 years, but it's depressing to think of a time as long as the whole development of civilization, spent hiding from the elements. . . .

"The snow level is constant now, and has been for some time, but the glaciers advance and the oceans are doubtless frozen over. No more rain is reported at all. I can't imagine that I ever disliked rain.

"Our video had to be shut down a year ago, so we are pretty much reduced to reminiscing for entertainment. All our stories have been embellished a lot, naturally. But sometimes I get depressed. Thousands of years of thought, civilization, and technology — all down the drain. Maybe in a billion years some spaceship will land and find new jungles here, sprung from seeds and spores that wait out the ice. . . .

"The S.A. experiments in some shelters aren't doing very well. We built more reliable suspended animation equipment a century ago. The Big Shelter reports that some of their systems have been operative for 30 years, but the typical breakdown occurs in only 20 years or less. I still think we'll solve the problems, and if our crew leaves here for the Big Shelter, four of us will stay behind in the cylinders. At least that's the current plan."

Cady stopped reading. She had put herself on record as a volunteer for the cylinders, but no choice had been made yet. The report of breakdowns was ter-

rifying; she hadn't known it was that bad. Keita was an
S.A. technician and had obviously kept the information
to himself.

Wondering why no one else had come into the
lounge, she went on to later sections of the journal and
found a passage that stopped her scanning eye:

"Few of us have opted for the old tradition of marriage here in the shelter, but I have thought about it
often over the years. Communal sharing has never
appealed to me, but since I'm afraid the others will
think I'm a throwback, I follow the custom just enough
to avoid suspicion. I would have chosen one woman as
my wife if she had ever indicated any interest, but
much as she obsesses my dreams, Cady remains indif-
ferent."

She reread the short passage several times, unable
to believe the words. Keita! Keita, whose lithe black
body and serious manner had always intrigued her. The
almost mystical look in his eyes when he spoke about
his heritage in the dim past. She had no idea! She herself had always disliked the whole idea of sexual community and had shied from it by burying herself in
work. If Keita had ever approached her in any personal
way, she would have assumed it was as fleeting and
physical an interest as those things had come to be. She
tried to think. Would it be too late now? Her body was
that of a 30-year-old woman, but her mind. . . . Still,
she felt deeply moved. How could she let him know?

Cady glanced at the wall chrono and jumped. Sol!
There was a community meeting this morning! No wonder no one had come in. Hastily she tucked the little
volume back into the cushions and ran from the room.

Two months later Keita and Cady were married.
Two years after that they argued successfully with the
selection team that both of them should go into the cylinders when the time came. Twenty years later they

exchanged a last embrace before stepping into their respective tubes.

"My beloved Keita is still suspended, so I carry on the journal he began over 3000 years ago. I was so relieved to discover I could still read his words, for I might have awakened without any memory. It's hard to believe the chrono, but its crystal mechanism was certainly expected to last and nothing seems wrong with it.

"My first sensation was terror. For a minute I was afraid to press the exit button, lest my cylinder door shouldn't open. But there wasn't much choice. Either I died quickly inside or took the risk.

"The air hit me like a hammer, and it took a few hours to get out and move my limbs. The shelter section in which I was embalmed seemed quite undisturbed, but the temperature is −40° C. I wanted to try the radio, but with the chrono's unbelievable message, there wasn't much point. Then I was struck by a terrible thought! The S.A. system had clearly failed, at least for my cylinder, and, long as it had lasted, it had drained too soon. The radio would be a distraction from looking at the other cylinders, which frightened me most of all. I finally made myself turn to them and, miraculously, everyone looked fine. It's easier to look at Morro and Tally than at Keita. I'm still fighting hard against the urge to bring him out, but the data still recording from outside indicate the glaciation is worse than ever.

"I must conserve what little strength I have to make these few notes. Then I can only hope the spare cylinder is functional—and that I can activate it. I'm afraid to start the cycle for a test until I'm inside.

"Let me just record here that I have been granted an extra 3058 years: the year is 5101. As I have no way to test the suspension formula, I must assume its still-liquid state is a good sign. Even if I could live off the remaining supplies, I have no wish to do so now. I have had the radio on scan these few hours, and there are no signals. I will join the others before I get too chilled to move—then freeze with the dream of my dear Keita. I hope we have a future together."

In shelter E-013-749 a piece of paper wafted to the floor as the intense cold arrested Cady's movements before she could reach the safety of the spare cylinder. As her emaciated body slid down the outside of the shining steel cylinder, her face filled with utter despair as her brain told her that the end was at hand.

The ice broke through the airlocks only 56 years later.

OUR SUN, OUR GALAXY, AND ICE AGES

CHAPTER SEVEN

Major ice ages—or rather the absence of them—have played a very important role in the evolution of life. Extended periods of warmth are needed to increase food production enough for the support of larger populations and the evolution of intelligent species, which in turn can develop a complex technology. Only during the past 11,000 years, since the glaciers last receded from the North American continent and most of Europe, did humans emerge from caves, start to live in villages, develop agriculture and finally technology up to the stage at which we now find ourselves, that is, on the frontiers of space. But how long can this mild state of affairs continue?

The cause of ice ages has long been a mystery. Recently two theories, both involving astronomical events, have received much publicity. One or the other is probably correct, and maybe both explain parts of the picture.

Despite the clockwork regularity with which the earth appears to move about the sun, it does in fact wander a little this way and that. Sometimes it is closer to the sun than its typical distance; at other times it is three million kilometers farther away. This uneven path is believed to produce large-scale temperature changes on the planet. If it happens to be summer in the northern hemisphere when the earth is a little closer than usual, then we experience very warm weather here. But if the earth moves a little farther from the sun, then the summers will get colder, as well as the winters and the overall effect will be to cool our planet down a little. This and several other peculiarities of the earth's orbital path around the sun are believed to produce cyclic patterns in climate that bring us repeated patterns of glaciations. This is part of the first theory for the cause of ice ages.

In addition to its variable distance from the sun (changes that cycle once in 90,000 to 100,000 years), the earth also wobbles on its axis, an effect known as *precession*. Right now our North Pole is located under the polestar, Polaris, but this has not always been so, nor will it always be. The earth's precession wobble goes through cycles lasting 26,000 years each. These are also expected to produce climate change, as the way the sun's heat reaches the earth changes with this cycle. The earth's axis also changes its tilt systematically in another way, *the change of obliquity*, with a cycle of 40,000 years or so. Combining these various cycles leads to predictions of planetary temperature changes that follow a regular, if complicated, pattern. There will be times when all the cycles act to produce a warming and other times when they all combine in just the right way to produce an ice age.

One can draw a parallel with the modern projection of biorhythms. Sometimes the biorhythm cycles are supposed to work together to produce a particularly good day, or at another time they allegedly combine to produce an off day. With the earth-orbit theory for ice ages, originally suggested by the Russian scientist Milankovitch in 1941 and only recently (1976) supported by good data on the actual occurrence of glaciations over the past 800,000 years, it is possible to make a firm prediction about the next ice age. Unlike the bases for biorhythm theory, the cycles in the earth's perigrinations are well measured.

The theory predicts that the time for the next glaciation is upon us already. It should take a few thousand years to get into full swing, then will last another 60,000 years. Do really have that to look forward to? It is likely that many decades will pass before scientists are absolutely sure whether

The galaxy known as M 51 shows a dramatic spiral pattern of luminous material. Dust lanes located along the inside edge of the spiral arms are clearly visible. Hale Observatories photo.

the theory is correct. And if it is, then it will take more time for people to take the prediction seriously, as is the way of human nature.

Even as astronomers and earth scientists battle over this theory, its tests and implications, a farther reaching theory for the cause of ice ages comes to our attention. It is possible that phenomena on the scale of our whole galaxy are producing the ice ages and hence are relevant to our existence on earth. Perhaps, as we said, a combination of the two theories may ultimately provide the correct explanation. Let's look at this second theory.

The apparent relationship between ice ages and the location of the sun in the Milky Way, as the sun travels between the other stars, is one of the most remarkable indications of how things out in space influence us here on earth. Ice ages may well be produced when the sun moves into dust clouds that drift between the stars in the spiral arms of our Milky Way Galaxy.

Many galaxies are flat disk-shaped objects with their youngest, brightest stars arrayed in spiral patterns. Not only the stars but also the gas (mostly hydrogen) and dust (of unknown composition) that lie between the stars are arranged in these spiral patterns. The term "spiral arms" suggests two armlike appendages neatly issuing from the center of a galaxy. And so they are in a small number of spiral galaxies, but very often more than two spiral arms can be seen. In some galaxies, the arms can be followed for several turns about the center of the galaxy; in others, the pattern may be very open, with the arms reaching far from the center of the galaxy without making even one complete turn. Other galaxies show no spiral pattern at all.

Astronomers believe the Milky Way is a spiral galaxy, a conclusion they reached only after carefully studying the distribution of stars, gas, and dust in the neighborhood of the sun. Because we are located inside one of these flattened, disk-shaped, spiral galaxies, it is very difficult to figure out what the whole galaxy looks like. We can never get outside it for a clear perspective of its shape, but enough data has now been accumulated to show that the Milky Way is a spiral galaxy.

But why should galaxies take on these spiral shapes? Not surprisingly, galaxies that show clear spiral structure are found to be rotating, and their rotation can quite easily be measured. The Milky Way itself rotates, with the sun and its environs moving about the galactic center at about 250 kilometers per second. That means it takes earth about 200 million years to move once all the way around the center of our galaxy. This is called a galactic year. Our planet was a very different place a galactic year ago, when dinosaurs were on their way to ascendency; even the skies were different. In a rotating galaxy with a spiral pattern, the spiral arms should wind up, much as a watch spring might. As a result, in a few galactic years we might expect the spiral arms to become so tightly wound that they disappear. But galaxies are found to be billions of years old, from studies of the ages of stars within them, and are thus many galactic years old despite their spiral structure. How can the spiral patterns persist so long?

The most credible theory about the maintenance of spiral arms suggests that there is an underlying wave pattern in the galaxy itself—a wave that looks to an outside observer like a spiral. This wave is manifested as density variations; that is, some parts of the galaxy are more tightly packed with matter than their immediate surroundings. This "density wave" moves around the center of the galaxy and through its stars, so that millions of years in the future the wave will have traveled way past any given star. To an outsider, that galaxy will look different, although it will still be a spiral-shaped galaxy.

But why do spiral density waves exist at all in galaxies? The solution may lie in the interactions *between* galaxies. Through computerized experiments in which hypothetical, *uniform* disks of matter pass by one another in space, it has been found that the matter in the imagined galaxies gets drawn out into spiral-like streamers, particularly at the edges. This might help to generate a spiral density wave that would then persist and continue to move about inside the galaxy. This spiral pattern would persist even after the two galaxies stopped interacting. In the case of our galaxy, two neighbors, the Magellanic Clouds (two nearby companion galaxies visible in the southern skies of earth and located about 160,000 light years away), may well have been the trigger for our density wave, since they appear to have passed close to the Milky Way several times during its history.

If spiral structure is really triggered by the interactions between galaxies, and if the correlation of ice ages with the passage of the sun through spiral arms is a valid one, then it appears that the terrestrial phenomenon of ice ages is related to the fact that another galaxy passed close to our own Milky Way billions of years ago.

As far as we in our orbit about the sun are concerned, the density wave in our galaxy is continually sweeping past the sun at a fairly slow rate (about 10 kilometers per second), and may trigger the process of star formation. The theory suggests that when a wave reaches any given part of the galaxy, the gas and dust there get piled up in clouds so dense that they are likely to collapse and form stars. The density wave therefore produces a wave of star formation; a spiral arm is a spiral-shaped region in which star formation has just occurred. The youngest stars, those most recently formed, are also the brightest stars in galaxies. Hence the spiral arms always appear as the brightest part.

In relating this phenomenon to ice ages, another fact is relevant. In other spiral galaxies, the great clouds of dust between the stars are located in streamers along the inside edge of the spiral arms. This is where the density wave has compressed the material the most. We can therefore assume that in our own galaxy the sun will sometimes be in regions of space where there are lots of dust clouds and at other times will lie in the space between spiral arms — in areas free of dust. When the earth and sun are inside a dust cloud they will, by their gravitational pull, sweep up some of this interstellar dust, much like a giant vacuum cleaner. The attracted matter can then fall into the sun, and it has been suggested that this might increase the sun's brightness. Such an increase in brightness would lead in turn to an effect on the earth's climate. This suggestion was made recently by the British astronomer W.H. McCrea.

The full explanation of severe climate changes, however, seems to involve not so much the brightening of the sun as the pollution of the earth's atmosphere by this dust. This suggestion arises when we compare the recurrence of galciations on earth with the sun's path in and out of dust clouds in space.

Major ice ages appear to recur about every 250 million years, each one lasting a few million years. This period is actually called an ice epoch, during which there are several of the distinct glaciations we normally picture as ice ages, separated by about 250,000 years. Each glaciation lasts about 50,000 years and the most recent one ended only about 11,000 years ago.

Professor McCrea showed how to estimate the time scales associated with the sun's passage through spiral arms, bearing in mind the density wave theory for the maintenance of these arms. His estimates lead to very similar time periods, coinciding nicely with known glaciations.

Since the speed of the spiral waves past the sun is known and the width of a typical spiral arm is also known, we can calculate how much time it takes for the sun to pass in and out of successive spiral arms. This time interval is a few hundred million years, like the time between ice epochs. As the dust clouds are confined to a relatively thin lane on the inside edge of the spiral arms, the time for the sun to traverse that part of each arm turns out to be a few million years. This estimate corresponds to the length of an ice epoch. We know that individual dust clouds in the streamers are separate physical bodies that come in typical sizes. One last quantity is needed: the number of these clouds that exists in any given volume of space. This also is roughly known. Armed with these facts, we can calculate that the sun typically spends about 50,000 years inside a given dust cloud and moves in and out of several such clouds during the time it is in the dusty part of the spiral arm.

The striking similarity between these numbers (the times spent in dust clouds and the glaciation times) suggests that the presence of the sun in a dust cloud might well be the cause for the ice ages. Following a series of glaciations, there would pass a few hundred million years during which there would be no ice ages at all, because then the sun would be back roaming between the spiral arms. The theory that each glaciation is produced when the sun passes through a distinct dust cloud thus begins to look credible.

Why should the immersion of the sun and earth in a dust cloud produce an ice age? The explanation concerns the effect that cosmic dust has on the earth's surface temperature if it drifts into our atmosphere. As we have mentioned before, the solar wind constantly sweeps past the earth but its harmful particles are kept at bay by the earth's magnetic field. In turn, the solar wind acts as a shield to protect us from some of the cosmic rays that are coming to us from outer space, and from the dust that is normally drifting between the stars.

The sun's wind may in fact blow all the way to the outskirts of the solar system before being halted by surrounding dust and gas. However, if the sun should wander into a fairly dusty region of space, the solar wind would find much more resistance to its progress and would be stopped well *before* it reached the earth, regardless of what was happening on the sun's surface. In this case, there would be no protective solar wind to prevent large quantities of dust from drifting down. The additional space dust in our atmosphere would absorb and reflect sunlight so that less solar heat could reach the surface of the earth. This dust pollution in the atmosphere might produce enough cooling of the planet to account for a true ice age.

If this is the reason our planet cools during an ice age, can we predict when the next one is due? Only if the locations of all dust clouds on the sun's path through space are known, and they are not. It is easier to examine very distant dust clouds than to try studying one that may be near us, so at present we have no way to know where they are until we are nearly into the next one. Perhaps some day the subject of weather prediction will include a study of our interstellar environment.

Note that when we refer to "dense dust clouds" we are not talking about anything we could easily see from earth. To our eyes, there would be no difference in the skies whether we were inside or outside one of these clouds. The amount of matter involved is extremely small by our usual standards. Very dense dust clouds might contain about 10 million particles per cubic centimeter as opposed to 100 or so for a typical cloud in space; yet it is these "typical" clouds that could trigger the ice ages. The very dense clouds are so few and far between that the sun would be unlikely to run into one. Even 10 million particles per cubic centimeter is a density of matter that is very much less than the best vacuum we can obtain on earth!

Support for the dust-cloud theory of ice ages comes from studies of lunar soil samples. Analysis of these samples, taken from a range of depths on the moon, shows that there have been periodic increases in the number of micrometeoroids (small solid particles from space) hitting the moon in years past. Several such increases have been revealed by carefully analyzing the character and size distribution of particles in the core samples. These increases appear to have occurred over periods separated by a few hundred million years. This suggests that every few hundred million years the sun and its planets pass through regions of space containing more matter than usual, a hypothesis consistent with their passage in and out of dust clouds. When the sun is in such a cloud, the dust not only falls onto the earth but also onto the moon and the other planets. (It is even possible, according to some, that if comets are in fact swept up as the sun moves about in the galaxy, the number of these visitors from space might increase when we are in dense dust clouds, producing a spectacular display in the heavens.)

Analysis of lunar soil samples isn't sufficiently detailed to permit scientists to correlate changes in meteoroid bombardment with individual glaciations on the earth. Nevertheless, the samples do hint at confirmation of the ice-age dust-cloud model.

Our entry into a real dust cloud could take several hundred to a thousand years. This would be very gradual in terms of any one generation, and succeeding generations would have lots of time to adjust to the colder temperatures. As time went by, they would also have to face the fact that the ice sheets were growing inexorably and extending toward tropical regions. Cities would be destroyed; but perhaps by the time this happened our technological society would have figured out a way to change the climate artificially so as to balance the effects of the dust cloud's mantle. Plans for such massive climate alteration are being considered now for our neighbor planet Mars, in case we should ever want to colonize there. At this time we don't know enough to control the climate here effectively; it is an enormously complex process.

Another, milder form of ice age has occurred within the historic past, from about 1640 to 1715. During those years, according to observers in Europe, the winters were unusually severe. The snows came earlier, piled higher, and stayed longer than usual. Glaciers began to advance again. Summers were short and cool, and food shortages were commonplace.

This kind of Little Ice Age, as that period came to be called, seems to depend entirely on the relationship between the earth and the sun. The sun has tremendous influence on our lives, of course. It provides the energy for photosynthesis in green plants, which in turn produces most of the oxygen we breathe. It gives heat and light, and it emits radio waves, ultraviolet, and x-rays. In addition, there is the steady stream of particles being blown out from the sun, the "solar wind." If we consider it an extension of the sun's outer layer, then we could say that the earth is continuously moving through the solar atmosphere.

Many aspects of the earth's climate and atmosphere appear to depend on radiations from the sun, as well as on the solar wind. Weather patterns all over our planet are somehow driven by heat from the sun. Scientists are slowly improving our understanding of these relationships.

There is now renewed interest in weather prediction based on what is occurring on the surface of the sun, where things are far from smooth. The presence of dark blotches, known as sunspots, has been observed for thousands of years. Although sunspots are now studied by means of special telescopes, ancient peoples occasionally noticed that at sunset, especially on evenings when the sun could be directly observed, the red disk was pocked by dark markings.

Now we know that there is a cycle in the annual number of sunspots that goes up and down once every 11 years. Earlier in this century, many earthly phenomena were thought to be related to the sunspot cycle. Not only the growth of trees, but the quality of wine vintages, the rabbit population, and even the Dow Jones index seemed to vary with the sunspot number. There were even those who believed that the ability of a car battery to retain its charge was correlated with sunspots. These correlations are regarded as doubtful at best, but renewed attention is now being given to sunspots and their relationship to weather.

There is some justification for supposing that sunspot activity is an important property of the sun. During times of maximum sunspot numbers, the sun is more "active," that is, there are more explosive events on its surface, such as flares and prominences. It is a well-known fact that at times of maximum sunspot activity the aurorae on earth are much more frequent and dramatic. The reason is that many more particles from the solar wind leak into the protective shield of the earth's magnetic field. Once they do so, they can come crashing down into our atmosphere along the field near the poles, where they produce the aurorae. There is some indication that auroral storms in the northeast Pacific trigger large-scale weather systems that determine the weather in the United States as they move eastward.

When there are lots of sunspots, the earth is constantly bombarded by more particles from the sun. The interaction of

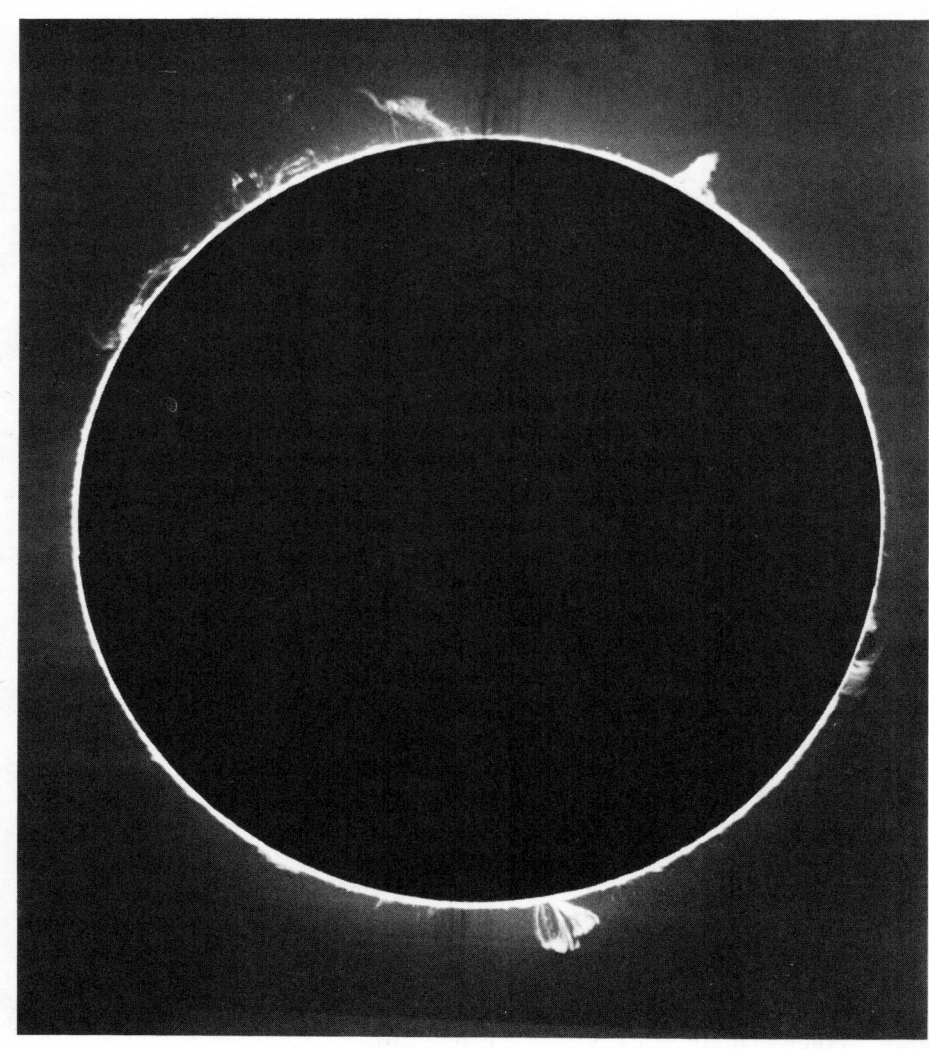

This dramatic view of the edge of the sun (its main disk being obscured) shows several prominences, arches of hot matter, that reach out from the surface, illustrating that the sun is far from being a smooth, quiet star. Hale Observatories photo.

these particles with the earth's field, and their subsequent leakage into the atmosphere, may trigger other climatological changes as well. Changes in climate, especially such dramatic events as droughts, are clearly a factor in the continued evolution and survival of countless life forms, humans included. One interesting discovery concerns droughts on the high plains of the United States, where droughts seem to recur every 22 years — soon after every second sunspot minimum. A drought did occur in 1976-1977, and during 1976 a very severe drought also hit England, Wales, and parts of Europe. A sunspot minimum occurred in 1975.

A very recent look at the cycle of solar activity suggests that the factor influencing climate may not be the 11-year cycle itself. There may be a more basic solar variable that determines both how the sunspot activity changes and how the climate is affected, but no one knows what that factor is.

A close study of sunspots, as recorded over the ages, shows they were not always present. During the Little Ice Age from about 1640 to 1715, virtually no sunspots were observed at all. There were scattered reports of perhaps one or two spots, but for the most part there were no sunspots visible. This phenomenon is called the Maunder Minimum, and it corresponded exactly with the Little Ice Age. Annual temperatures were then on the average of about one degree below normal. That doesn't sound like much when compared to typical ranges that we experience every day, but when averaged over any one year such a temperature drop means a lot. A major ice age results when the average temperature drops only 5 to 10 degrees!

During the Little Ice Age crops suffered; rivers never frozen before froze over and dozens of towns at high northern locations were abandoned. People suffered greatly, but our knowledge of just how much is limited. In the modern era, with the present massive populations on earth so susceptible to famine and with our critical interdependence on one another for food supplies, the consequence of another Little Ice Age would be nearly catastrophic.

Careful examination of the sun and climate have been possible for no more than 350 years, so indirect arguments have to be used to figure out how the sun behaved in earlier times. Some clues can be derived from historical records. For example, during the Maunder Minimum (first recognized at the beginning of this century, but mostly ignored by astronomers until 1976) there were virtually no recorded reports of sunspots, although close examination of historical records left by famous astronomers shows that they always wrote descriptions of observed sunspots. These were often in the form of papers published in early prestigious journals. Nowadays no one gets the slightest bit excited about another sunspot, so they must have been rare events indeed. In addition, there are very few records of aurorae during this time, as would be expected if there were no sunspot activity to generate the solar storms that produce aurorae.

The sunspot cycle may even be a recent manifestation of the sun. Examination of reports on solar activity as far back as 5000 years ago suggests that a sunspot cycle is not the rule. It might even be the exception. Climatic changes may well be related to much slower events on the sun which are not yet understood. To identify these changes historically would be extremely difficult. Problem enough to establish causes of short-term climate changes!

One might expect that such correlations between climate and sunspots would be easy to discover and that it would be a simple matter to figure out an explanation. Unfortunately scientists have neither enough information on the weather patterns over the planet nor enough data on the solar cycle — nor do they find it easy to sort out which climate effects are important for them to consider. One thing is clear, however: during the Maunder Minimum of virtually no sunspot activity, parts of the earth (certainly Europe) endured a very cold spell for some 70 years.

If we consider again the role of the solar wind, we may be able to frame a working hypothesis that connects these two observations. Under normal circumstances the steady

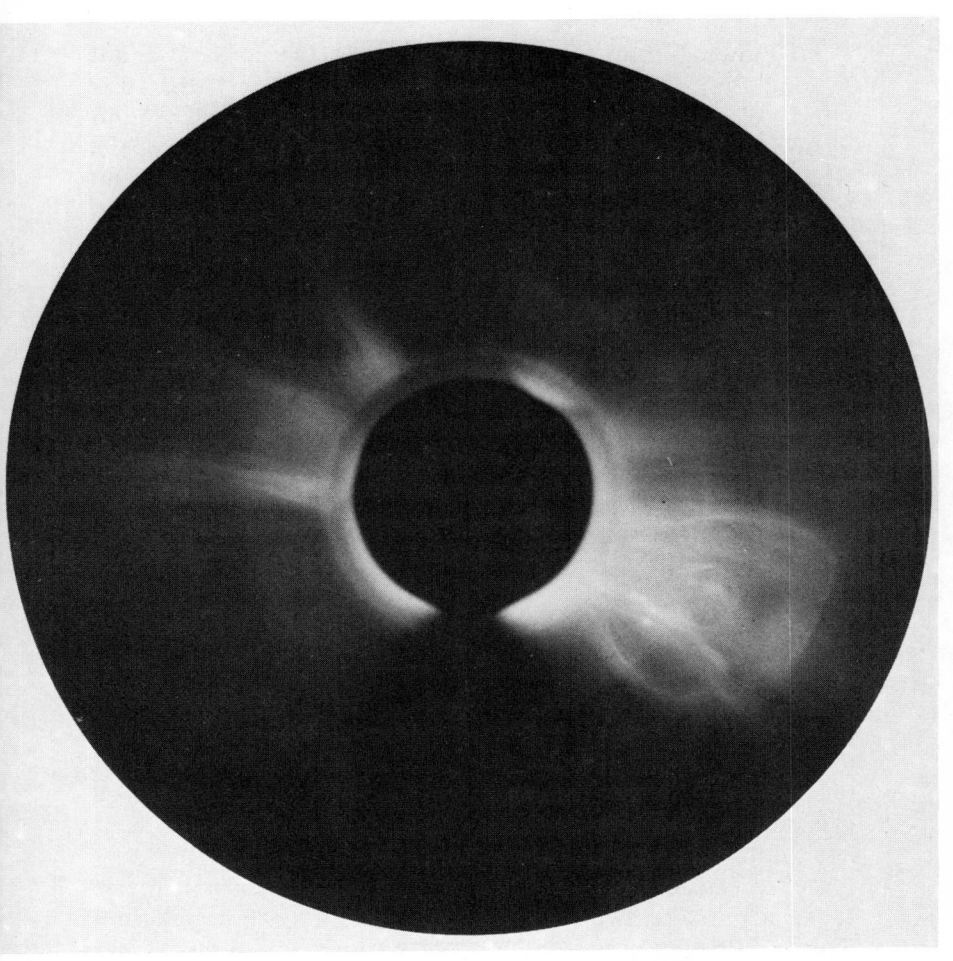

This view of the sun taken from Skylab shows not only the corona (atmosphere) of the sun radiating outward, but also a loop of matter moving out through the corona. Now called transients, over 100 such events were photographed from Skylab. They had never been seen before from earth, with a possible exception during a total solar eclipse in 1860. During the Little Ice Age the sun did not appear to manifest a corona. High Altitude Observatory and National Astronautics and Space Administration.

wind blowing from the sun sweeps past the earth and out beyond the orbits of Mars and even Jupiter. Recall that this wind holds at bay the dust particles that fill space all around us and all of space in our galaxy. Now imagine what happens when the solar wind drops a little. In that case it can no longer push as hard against surrounding interstellar dust. Then this dust is free to fall down into the earth's atmosphere, reducing the amount of heat and light that can reach the surface. If the sun were to lower or turn off the wind for a while, we might well find enough dust in our atmosphere to change the surface temperature.

It seems credible that in the Little Ice Age enough dust accumulated to drop the temperature about one degree in 70 years. Then the sun turned its wind on again and the dust was swept back out away from the vicinity of earth, raising temperatures to warm us again.

If a cooling of the earth is produced in this way (and at this time this is pure speculation) during each quiet phase in the sun's year-to-year variation, then the next Little Ice Age should happen as soon as the sunspot cycle dies out again. No one knows why the sunspots disappeared for decades at a time nor when this might happen again. But obviously if it happened before, it can do so again. And when it does we may all feel the consequences. Whatever the real explanation for the Little Ice Age may be, astronomers are keeping a close eye on the sunspot cycle. Currently it is behaving more or less "normally."

We earthlings appear very vulnerable to cosmic forces, especially in the twentieth century with its overpopulation. Even a Little Ice Age, which could happen any time, would cause havoc on our planet in a few decades. The greatest ice-age cataclysm would result if the two major processes (orbital and galactic) for cooling the planet were to act together. If the earth's orbit moved just a little farther away from the sun at the same time that we entered a galactic dust cloud, no one could even guess what would happen. And, of course, we have no way of predicting whether those two events might happen simultaneously.

One thing, however, appears certain. If the orbital theory is correct, we have very cold times ahead. Our civilization, which includes all the peoples on earth, will either mobilize to change the earth's climate, or will migrate to warmer places and try to stick it out.

IMPACT

CHAPTER EIGHT

Abby Morgan dabbed the snowy linen napkin carefully to her lips, sighed and settled back into the soft velvet cushioning of her chair. She watched Marlene's back receding from their table toward the powder room on the other side of the elegant restaurant they had chosen for their escapade. When the door closed behind her friend's slim form, Abby sighed again and gazed up into the myriad lights of the chandelier overhead.

She was pleasantly tipsy now, and no longer anxious. Some day she might even be able to eat in a place like this and pay for it. Some day. For now, she reflected, all the days she and Marlene had spent pushing around endless piles of heavy unaesthetic dishes in a luncheonette surely entitled them to one gourmet adventure. It was after a particularly frustrating day of sullen customers that Marlene, sprawled across one of the two improvised bed/couches in their one-room efficiency, had hatched today's plot.

"Come on, Abby," she had argued, "some day

we'll turn up real acting jobs. Why not a little practice in the meantime?"

Abby had agreed on condition that Marlene would do the special bit of acting called for when the check came. She had gotten into the spirit of things by dressing impeccably in the Givenchy sweater and skirt she had found in Boston's famous Filene's basement just before she came West last year. The little symbol embroidered on the shoulder was visible enough to give her confidence. Givenchy was Givenchy, and the rest was just a matter of exuding an air of wealth that would belie the bargain table. This was simply a preview of "some day," she told herself, when there would be more than occasional bit parts, when persistence and fortune would blend in stardom.

Stardom. Her name spelled in lights. With a slightly woozy mind, she floated into the tiny stars of the chandelier above her. Miniature bulbs were held ingeniously in space by nearly invisible wires, like the spray of a giant sparkler frozen in space, or a galaxy of real stars — stars for Abby.

Her reverie was broken by Marlene's return. Sipping the last of their liqueurs, the conspirators discussed the Clos de la Roche '72, the delicate paté and the superb grouse until the waiter approached with an ingratiating smile.

"Shall we put that on your bill, madame?"

"Yes, of course!" Marlene did not waste a second, scribbling something that could be taken for a name on the bill and adding a hefty tip for good luck as they left the table, before the waiter could return for the forged document.

The doorman bowed and scraped, Marlene having the presence of mind to pay him a sizeable tip as the final gesture in their controlled exit — a tip beyond the price of a meal in their more usual haunts. Abby began to berate her for the extravagance as they quickly rounded a corner, but then both of them dissolved in giggles of mutual congratulations.

Three blocks further into safety, they separated. Abby wanted to check a little theatre not far away where the play with ner new bit part would open in a week. Against odds, she prayed her name might appear on the playbill. Marlene caught a bus back home, since she expected an important call.

The twilight cast an eerie glow through the streets of Los Angeles and onto the trees and grass of the little park Abby passed. The touch of intense green, she later recalled, was a pleasant contrast to the enormous buildings looming above her. They weighed heavily on concrete floats (she had read) that provided protection against earthquakes. They wouldn't split unless a fault passed right through their basements. Thus, her last thought before the blast was of security against earthquakes, and her last image a sudden sharpening of shadows etched on the ground—intense dramatic shapes cast from the buildings, from lamp posts, from herself as the sky flared.

All over the United States people had watched the meteor shower grow. They had observed the unusual fiery tails curiously, but not too anxiously. Suddenly a great ball of light flashed across the skies and disappeared below the horizon. From Phoenix it seemed to go down in the southwest. In San Francisco, the flash lit up the southern sky. San Diego was bathed in a blinding glow from the north. Countless eyewitnesses saw the fireball itself, so bright that the entire western USA was lit as if by full daylight.

Some of the inhabitants of Los Angeles saw it appear overhead just before the blast shattered their city. It struck near the edge of the ocean at Santa Mon-

Earthbound Angel. Etching. Deep in the interior of the earth motions of molten material give rise to the magnetic field that protects us from matter constantly spewed out by the sun.

ica, and disaster spread from there. The explosion split people's eardrums miles away, tearing across the Los Angeles basin and ricocheting off the mountains. Its sound was clearly heard in Arizona and northern Mexico. San Franciscans heard it and responded instantly with cries of "earthquake!" Those that ran into the streets saved their lives, for only minutes later the quake did hit.

In southern California, the ground ripped open along 500 miles of the San Andreas fault. The enormous seismic energy of the meteorite's collision triggered the fault, already poised on the verge of catastrophe. The combined upheaval flattened most of the buildings in the Los Angeles basin. Abby Morgan was one of the half dozen survivors who recovered consciousness and staggered to their feet in central Los Angeles. She took a few dizzy steps, thinking that surely she would not have to worry about the restaurant bill now. This incongruous thought kept passing through her mind as she stumbled in the near total darkness. Slowly, the horror of her situation struck her. There were no lights, no moving vehicles. Turning, she saw a glow and heard the crackling of distant fires. The silhouette of a distant building moved before her eyes and slowly toppled over.

Was she delirious? Had she drunk that much? Tears of panic rose as she tried to run, but then the earth heaved beneath her feet and she threw herself down on the ground in anguish.

The world was coming to an end! Abby looked around wildly as another building toppled, and she clawed at the pavement for some vestige of security. She was acutely aware of distant figures trying to scramble away from tumbling debris, heard moans and screams, but was unable to call out herself. She lost consciousness a second time.

Dimly, as from a great distance, she became aware that someone was pushing at her limp form. Coming out of her daze, she could just make out the image of a little girl, kneeling beside her and sobbing.

"Mommy . . . my mommy . . . and my daddy. Please. I can't find . . . dark. Please, lady. Please help. . . ."

"Oh, God," Abby gasped, throwing her arms around the little girl and holding her so tight that the child struggled to free herself.

"Just please . . . please help me find. . . ."

Abby forced her attention into focus. She was the adult. She had to think clearly. She was not drunk; she was in the middle of a war. It had to be an enemy attack, a nuclear missile. China? Russia? No, she must not try to figure that out now; she had to help this sobbing child, to find other people.

She managed a stoic smile in the dark and asked:

"What's your name? Tell me your name and I'll help you look for your mommy and daddy. I'm Abby."

"Diane," came the tremulous reply.

"O.K., then, Diane, I think we'd better get away from here right now, then we'll go and find them."

Struggling upright, she became aware that one of her shoes was missing. After groping around for it briefly, she abandoned the effort and pulled off the remaining shoe. Her legs felt too unsteady for heels anyway. Suddenly Abby was grateful to find herself uninjured. She took the child's hand and began to seek a way out of the desolate area surrounding them. The distant flames licking into the sky created a glow and dancing shadows all around them. But their path was blocked in every direction they tried by heaps of rubble like movie scenes of wartorn Europe or Hiroshima. Ominous rumblings and the crackling of fire were the only sounds relieving the now inhuman silence.

Abby wanted to scream but couldn't bring the sound to her throat. Was she protecting the child against further panic? Was there something wrong with her throat? Or did she somehow know that shouting would bring no response? Diane, too, was silent; Abby wondered if she might be in a state of shock.

She pulled Diane down beside her and sat in the middle of the street. It was a wide street, an open space that might offer at least some protection from falling rubble and fire. Anyway, there was nothing else to do.

"Let's wait here for a moment," she whispered, holding the now unresisting form close to her.

"What happened? Where is my mommy?" Diane began to sob again, more weakly now, as exhaustion overcame her. There were no answers to give, so Abby just held the little girl, patting and soothing—giving comfort as much to herself as to the gradually relaxing child.

Huddling in the open air, Abby tried to assess the situation. She recalled the sharp shadows, the strangely brilliant sky as she lost consciousness the first time. She tried to recall the sequence of events. There was no explosion in her memory, only the heaving earth as she tried to run later. Earthquake! That must be what happened. Not a bomb, but an earthquake! But why the flash of light? Oh God, and why no other living beings? If it was an earthquake, she was right to stay out in an open place. There could be more aftershocks. Maybe rescue missions would be coming soon. Did Marlene get home? *Was* there a home any more . . . a Marlene to worry about her? She did not even know what direction to look in for home.

A thunderous roar in the distance announced another shock and several buildings nearby collapsed as whole walls tore loose. A series of explosions punctuated the slower thunder rolls of falling masonry.

Abby then noticed a rising moon struggling through the clouds and smoke in the thickening atmosphere. She took her arms carefully from around the sleeping Diane and stood up shakily. Should they try to move on? In which direction? She picked her way carefully around jagged pieces of concrete and plaster, coughing in the dust and smoke that rose from the gaping buildings. For the first time, she became aware of bodies, human forms trapped and twisted in the piles of wreckage. Nothing moved. There were no human sounds. Another wave of fear overcame her and she picked her way back to Diane as quickly as she could.

There was nothing to do but pray and wait for daylight. Surely the rescue missions would come by then. She would scream then, when this dark nightmare lifted a little.

It was just about dawn when Abby was roused out of her fitful dozing by the sound of a whirring engine overhead. She sprang to her feet and began to shout wildly, flailing her arms and leaping up and down. The helicopter came into sight, and on a second turn over the two frantic figures, it began to descend. Blowing clouds of dust in all directions, the rotor never stopped as the two survivors were scooped into the copter by helping hands.

The sun rose red as Abby and Diane were lifted into the air. Speechless and wide-eyed they stared at the smoldering disaster below, more and more of it coming into view. They had been retrieved from the edge of a totally devastated area. A crater, miles across, lay where West Los Angeles had been. Nothing remained there. Hundreds of smaller craters were scattered around the main one, some of them very near the spot from which they had been rescued. Tails of smoke swirled in all directions. Abby froze as the full extent of the devastation struck her. Somewhere down there she had had a home. Somewhere down there all her friends now lay dead . . .

Gradually she became aware that someone was talking to her.

"What . . . was . . . it?" she stammered.

"Are you O.K., lady? Is the kid O.K.?"

"Yes, yes," she answered, "but what *was* it? Is it war?"

"Meteorite," her rescuer answered loudly, over the whirring of the rotors. He continued to search the ground below for other moving figures as he explained.

"A giant meteorite, busted into pieces. Then earth-quakes. You sure are lucky; not much left down there. Lot of explosions; gas mains and stuff. Didn't think we'd find anyone here. Don't see any others. Ground crews will get around soon; we're just looking for the ones on their feet."

Abby tore her eyes from the scene and put her arms around Diane's shoulders. The child had not uttered a sound and seemed so small and helpless. Her parents would probably never be found. At least it seemed hopeless now. Abby wanted to ask where they were going, what was left, but it didn't seem to matter. Instead she shut her eyes against the grim sight outside her safe bubble and tried to get a hold on her own situation.

Her parents had been killed in a different kind of collision years ago. She had no siblings and had lost contact with her old East Coast friends, thinking of nothing but her career—not even of marriage. There was really no one but Marlene. And now—? She thought again of yesterday. Ironic! If she and Marlene had not played their charade in this part of the city, both of them would be dead now. Lives snuffed out, careers snuffed out, maybe millions of them. She wouldn't be going back to that grimy luncheonette any more; she also wouldn't be thinking of stardom for a while. She flashed back to her reverie beneath the chandelier at the restaurant. Saw again that galaxy of tiny stars. She had seen herself only yesterday among human "stars"; now, from somewhere out in the sky, had come this fireball, taking all she had hoped for, making the future unbearably lonely and unknowable.

Diane shivered in her arms. Maybe not entirely unknowable. This little girl had survived with her. Suddenly Abby needed her as she had never needed anyone before.

CRATERS EVERYWHERE

CHAPTER NINE

An alien visitor to the inner planets of the solar system—Mercury, Venus, Earth, and Mars, as well as the moons of Earth and Mars—would notice, above all, the craters that are found on all of them. With the naked eye, we can see some of our moon's very old craters as dark areas we call *maria*. Photographs returned to Earth from spacecraft visiting Mercury and Mars have revealed Mercury as heavily cratered and Mars as somewhat less so. What were the events producing so many craters and do they still occur?

Some of the craters are now known to be volcanic in origin, but the vast majority are produced by explosions caused when some piece of interplanetary debris hits the surface of a larger body with incredible force. This debris has been tumbling around in space between the planets since the formation of the solar system. In the earliest days after the birth of the solar system, 4.7 billion years ago, these stones, rocks and boulders in space constantly struck the newly formed planets, as well as their moons, with great regularity. Some pieces of cosmic debris were of enormous size. When they hit the moon, for example, they produced gigantic craters hundreds of kilometers across, visible to us from earth. Since there is neither atmosphere nor water on the moon, there is no erosion; therefore, since those old craters have remained untouched except by further collisions with space debris, they are still visible to us today.

In the early solar system there must have been a complete range of sizes in the objects orbiting around the sun. It is believed that the most massive objects, say the size of a small moon, were least numerous, and smaller objects, all the way down to dust particle size, were more numerous. If so, the planets, once formed, experienced relatively few collisions with large objects, but countless collisions with smaller objects.

As the solar system aged, space between the planets got swept clear of this debris, although even now, between the orbits of Mars and Jupiter, there are thousands of objects ranging in diameter from a kilometer or two to hundreds of kilometers. These objects, called asteroids, probably never gathered to form a planet.

Some of the matter that even today impinges on earth originates in the asteroid belt. When a piece of matter from space comes hurtling toward the earth, we often see a meteor or "shooting star." Most of these objects are very small and burn up rather quickly once they hit our protective atmosphere. As it burns, a meteor gives off an enormous amount of light and can typically be seen several hundred kilometers away. Tiny meteors occur very frequently; on a clear night you should be able to see many meteors in any hour. But you will see a fireball only once in several years, if you are lucky.

A meteorite is matter from space that survives its passage through the atmosphere and crashes to the ground. It is usually composed of stone or iron or a mixture of both. The meteorite also produces a bright incandescent trail through the air called a *fireball,* or *bolide.* Observers near a fireball can often hear an explosion as it is shattered — usually in the air some distance above the ground — by its violent heat. When a fireball is well observed by many people, it is possible to pinpoint where it came from and where it landed. On the average, one such landing per year is tracked down somewhere in the United States. Usually the meteorite is recovered and taken to a museum. The second largest meteorite ever recovered in the United States was found in California in 1977.

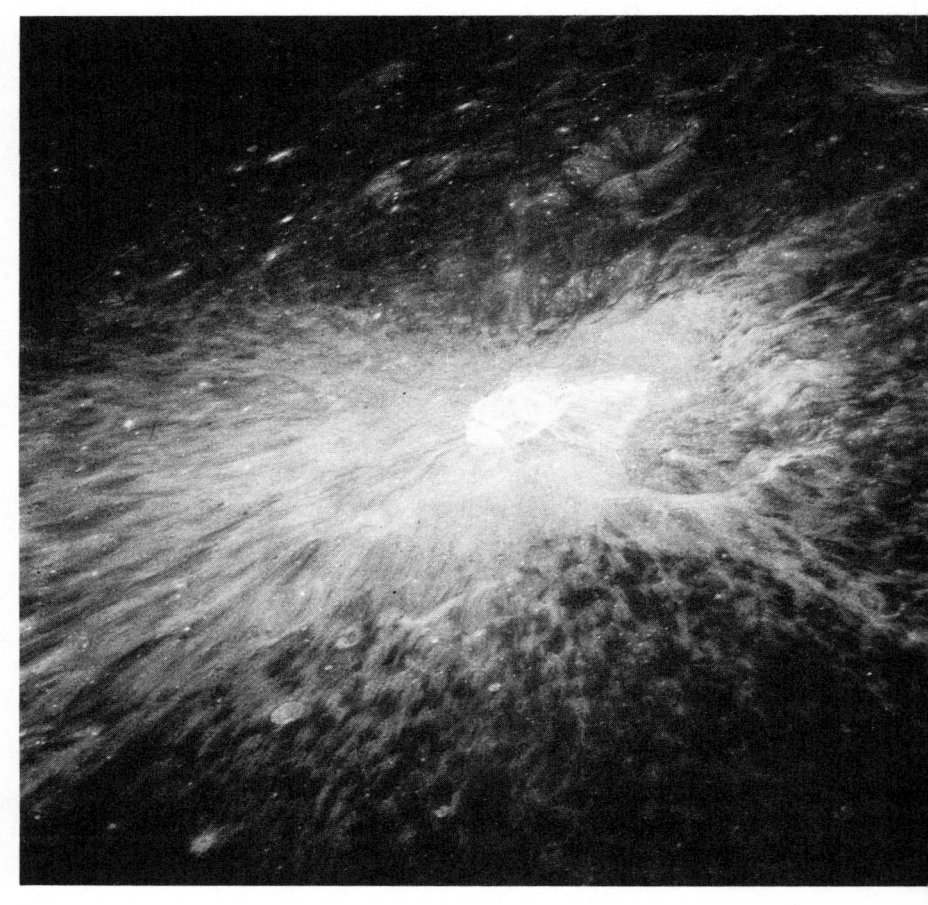

A lunar crater as photographed from a lunar orbiting spacecraft showing the dramatic splash zone where matter was thrown out onto the surrounding surface. The majority of lunar craters were produced by impacts with meteorites during the early years of the solar system's existence. NASA photo.

Meteors are very different from meteorites. They are much smaller and appear to originate in a different part of space. While meteorites generally come from the asteroid belt, meteors seem to come from beyond the solar system, from the same regions of space that give rise to the comets, those fiery visitors that sometimes orbit the sun closely and produce dramatic, luminous tails that can be seen with the naked eye.

Recently the rocks in meteorites have been chemically tested very soon after they landed to assess their chemical composition and their age. From such measurements come estimates that they formed in the early solar system about 4.7 billion years ago.

The largest meteorites, weighing many hundreds of tons, fall through the air quite unimpeded by the atmosphere. They can strike the earth with staggering force. Typical meteorites can range in weight from a fraction of a kilogram to several tons. In one case on record, a woman in Alabama was struck in the hip by a five-kilogram meteorite that left her severely bruised and frightened (fortunately, it hit her on the first bounce).

The most famous meteorite collision with earth left its scar in the Arizona desert. The Barringer Meteorite Crater, near Canyon Diablo, is a little more than a kilometer across and about 150 meters deep. Believed to be 22,000 years old, the crater is surrounded by thousands of fragments of the original meteorite, which must have exploded on impact. Its main remnant is still believed to be buried somewhere below the crater.

A 100-ton meteor is expected to produce an explosion equivalent to a million tons (one megaton) of TNT; that is, equivalent to a small atomic bomb. It is estimated that the Barringer crater was produced by an explosion equivalent to three megatons of TNT. However, there are even larger craters found on earth, some as much as 24 kilometers across, which indicate explosions of up to 2000 megatons of force, larger than any explosion ever set off by humans. Collisions of this magnitude are still expected to occur about once in a million years, although we need not lose too much sleep over this happening soon. Of course, if such a meteorite were suddenly to strike a populated region of the earth without warning, we would have a disaster scene unprecedented in history, and more terrible than we can imagine.

The other planets and their moons, as well as our moon, show the scars of such catastrophic collisions in earlier times when they were much more frequent. Most of the meteorite craters on earth have long since been eroded away or churned under by the processes that shift continents and build mountains. However, we should bear in mind that the existence of meteorite craters on earth was only established in this century, when the Barringer crater was recognized for what it is. In 1927, clear evidence of a second meteorite crater was found near Odessa, Texas, where there is a crater 150 meters across and 5 meters deep. In Australia some 13 craters, covering an area about 470 meters across, were found near Henbury Cattle Station in the desert. Such clusters of craters have been found in several other places since, and are clearly due to a meteor shower consisting of unusually large lumps of rock, each of which dug its own hole.

The twentieth century has brought several dramatic visitations from space. At 7:17 A.M. local time, June 30, 1908, a searingly bright trail was seen crossing the sky in north central Siberia. Eye and ear witnesses reported the long shining trail and a series of sounds like explosions of gunfire, or like tons of stones being dropped. The object's fall was seen by people more than 1500 kilometers away, and the explosion was heard 1250 kilometers away. In fact, sensitive atmospheric instruments still picked up the sound wave even after it had rounded the earth twice. Whatever the object was, it fell far from any inhabited area. The nearest village in which a newspaper recounted the Tunguska valley event was 200 kilometers away. Peasants living nearer the explosion had dramatic tales to tell, but strangely they were never gathered until 1928. Then, in quick succession, three expeditions of scientists searched for the site of the explosion, which was located and thoroughly explored. Trees over an enormous area were found to be flattened and lined up radially, pointing away from a spot that must have been the center of the blast. Trees up to 20 miles away were seared and damaged, while those closer in were actually flattened.

Eyewitness accounts gathered from the local people are fascinating to read. One peasant reported: "At that time I was ploughing my land at Narodima. When I sat down to have my breakfast beside my plough, I heard sudden bangs, as if from gunfire. My horse fell on its knees. From the north side above the forest a flame shot up. I thought the enemy was firing, since at this time there was talk of war. Then I saw that the fir forest had been bent over by the wind and I thought of a hurricane. I seized hold of my plough with both hands, so that it would not be carried away. The wind was so strong that it carried off some of the topsoil from the surface of the ground, and then the hurricane drove a wall of water up the Angara [a river]. I saw it all quite clearly because my land was on a hillside."* This eyewitness was located about 240 kilometers from the site of impact!

As far as we know, only reindeer died in the blast, but several people were inconvenienced, as in this report: "I am a tanner, and in the summer about 8:00 A.M. I was washing wool on the bank of the river Kan. Suddenly a noise like the fluttering of the wings of a frightened bird was heard coming from a southeasterly direction from the village Antsyr, and a kind of swell came up the river. After this came a single sharp bang so loud that one of the workmen, Vlasov, fell into the water."† This occurred over 160 kilometers away. At least one forest dweller suffered loss of hearing and was deprived of speech for a long time, according to his brother's story.

For two nights after this cosmic visitor collided with our planet the skies were very bright throughout the night. Even in Scotland it was reported that newsprint could be read at midnight. This widespread and persistent glow was a mys-

*E.L. Krinov, *Giant Meteorites* (Elmsford, N.Y.: Pergamon Press, 1966).
†Ibid.

terious manifestation of the interaction between the matter in the meteorite and the atmosphere.

Expeditions that visited the site in 1958 found evidence that the direct blast wave of the explosion reached nearly 80 kilometers from the center. The expeditions found traces of meteoric dust in the soils, but no solid meteorite material was ever found and it is believed that the Tunguska event was actually a small comet that struck earth after entering the solar system. Its total mass may have been some 40,000 tons, and it seems to have fragmented totally just above the ground, raining dust and small particles all around. The comet hypothesis accounts more easily for the bright nature of the trail it left as it streaked through the atmosphere.

It is natural to consider what would have happened if the Tunguska meteorite/comet had hit a populated area. A small city would have been totally devastated and countless people killed. The time and place of these events cannot be predicted.

In 1947, 200 miles north of Vladivostok, another incredibly bright fireball was seen. It was so bright that its radiation cast shadows even though there was full sunlight at the time. Eyewitnesses reported that it was about the size of the moon and occasionally brighter than the sun. The point of impact, quickly found, revealed no fewer than 106 small craters, ranging from about a meter up to more than 20 meters in diameter. An explosion had totally shattered the original meteorite, again just before impact.

Despite the continuing bombardment of earth, there is no authenticated record of anyone ever having been killed. However, a report dating back to 616 B.C. claims that several chariots and 10 men were destroyed by a meteorite. In recent times several people have had close calls, although the Alabama lady remains the only person known to have been struck.

One of the most dramatic recent sightings was reported from the Kirin Province of China, where, on March 8, 1976, a meteorite shower occurred. A red fireball as large as the full moon left a smoky trail, breaking up several times as it flashed across the sky. The closest eyewitnesses on record were three traveling peasants and three children playing outside the village of "Kaoshan No. 10 Production Brigade of Huapichang Commune of Yungchi county west of Kirin city." The main segment of the meteorite dug a hole some six meters deep and two meters wide, characteristic of smaller meteorites that have been slowed down by the atmosphere. These tend to produce clearcut holes rather than craters. The meteorite shower that accompanied the fall covered an enormous area of countryside, nearly 80 kilometers wide in one direction and 8 kilometers in the other.

Who knows when an unwelcome visitor from space may next come crashing down on us? The chances of a meteorite hitting a major city are of course very small, but the Tunguska event of 1908 is there to remind us of our vulnerability. Had the object struck our revolving earth five hours later than it did, the city of St. Petersburg would have suffered a direct hit, according to some who have carefully analyzed that event.

No one yet understands what caused the violent explosion in the Tunguska valley, although the comet theory appears most reasonable. One farfetched theory proposed to explain the Tunguska event is that it was due to a nuclear weapon, even though it occurred back in 1908. Such a nuclear explosion would necessarily have had an alien origin, since no one on earth had even dreamed of nuclear weapons in 1908. This explanation has been disproved, however, because such a nuclear explosion would have produced chemical changes in the earth's atmosphere that would be detectable in the chemical analysis of tree rings. The carbon content of tree-ring samples shows that nothing like a nuclear explosion could have occurred in 1908.

Another theory for the mysterious Tunguska event suggests that a miniature black hole (see Chapter 11) may have plunged into the earth in Siberia, probably to emerge somewhere in the Atlantic Ocean. However, some scientists have opposed such a theory, with convincing arguments based on the expected interaction between a black hole and earth. Suffice it to say that the predictions of such a theory do not accord with the observed effects of the mysterious collision in Tunguska.

While the chances of hitting a city are very small, a giant meteorite crashing even into the ocean would have serious effects on life. The splash produced by a 100,000-ton rock crashing into the ocean would be almost beyond imagination. It would produce unprecedented tidal waves that could, in turn, produce incalculable devastation along coast lines, even along such large oceans as the Atlantic or the Pacific.

The comfort we can derive from our present knowledge of conditions in the solar system is based on the observation that things are relatively quiet now. Most large bodies that were likely to hit the planets have already done so. Yet even now it is expected that a meteorite capable of producing a crater more than a few meters across will hit the earth every 1000 years and produce a blast equivalent to 1000 tons of TNT. Every 40,000 years a Barringer-size crater might be produced, the blast being equivalent to three megatons of TNT. The largest meteorites that might still be expected to strike the earth, producing craters greater than 13 kilometers across, are expected once in a million years. A meteorite of this size would produce a blast equivalent to 2000 megatons of TNT, far greater than anything we humans could produce even with our most destructive bombs.

The shocks produced in the earth's crust by a large meteorite would be capable of triggering earthquakes, especially if it were to fall in an earthquake belt. A fault like the San Andreas, already greatly stressed, might be triggered to violent motion by the impact of a large meteorite in or near California. (Note that the "Jupiter effect," a hypothesis suggesting that the San Andreas fault would be triggered when the planets all line up on the same side of the sun, is not discussed in this book because the hypothesis has little support in the scientific community.)

The emergence of life on earth occurred only after the main meteorite bombardment was over. For the past billion

The northern limb of the planet Mercury, as photographed by Mariner 10. Being unprotected by an atmosphere, and with no erosion taking place on its surface, Mercury is still pitted with craters, revealing the enormous numbers of impacts it has been subjected to since birth. NASA photograph.

years or so, things have been relatively quiet. Possibly, large-scale stable societies could emerge only after the bombardment ceased. It is hard to imagine living in a civilized society that is continually blasted by giant rocks from space. Whether the next giant meteorite will produce great destruction is entirely unknown, but one thing is certain: there are many travelers between the planets that could come crashing down unexpectedly at any time, although most of them are no greater than pea-size. Watch the fiery meteors you can see any night and reflect on the fact that they are just tiny remnants of the enormous amounts of matter that used to fill interplanetary space and that still shower all the planets and moons indiscriminately.

Several tons of meteor material, mostly in the form of small particles, drop to earth every day from isolated "shooting stars." At regular intervals meteors also occur in spectacular showers, such as the Perseid shower that recurs every year on August 12, when the earth crosses the path of a comet that visited the solar system in 1862. Some years a given meteor shower can be particularly dramatic, with up to thousands of meteors visible every hour. For example, every 33 years the Leonid shower, on November 16, is of this dramatic nature, with as many as 140 meteors a second having been recorded. Its next dramatic appearance is scheduled for 1999.

The meteorites that actually fall to earth interest scientists more than shooting stars. Analysis of their composition tells us something about conditions in the early solar system.

When a good fall is recovered, scientists cut the meteorites open to reach the undisturbed interior. There they can find the most accurate traces of what the solar system used to be like, or at least that part of the solar system in which the meteor originated.

The first step is to measure the age of the meteorite by assessing the amount of various elements in the object. Since certain radioactive chemical processes produce rare elements such as thorium and rubidium in particular amounts, and the amounts present change with time, it is possible, by measuring the ratio of these elements in the meteorite, to estimate how old it must be. Typical ages seem to be around 4.7 billion years, which is also the age estimated for assayed moon rocks. This, then, must be the age of the solar system. Now, nearly five billion years after their formation, some of these rocks or lumps of iron still come crashing down on us.

Another fascinating discovery was made in recent meteorite analysis. Using the techniques developed to study lunar rock samples, scientists found that meteorites contain organic molecules, so basic to life on earth. In fact, important amino acids fundamental to all the life we know have been found in meteorites. Amino acids have not been found on the moon or Mars. In addition, some of the amino acids in meteorites are quite different from any found on earth. This difference is described roughly in terms of the way the atoms are structured in the molecule. Amino acids can form in a given pattern and also in the mirror image of this pattern. On earth only one type is found, without its mirror image. However, both types have been found in meteorites. This proves that somewhere in the very early solar system molecule-building of the sort necessary for life was already taking place, but that it formed two classes of these important molecules. The fact that the earth naturally has only one of these types of amino acids seems to be a fluke of nature; it could equally well have been the other way around, and life would probably have been different. Why earth has one and not the other is not known.

Whatever it was like in the solar system several billion years ago, we do not know all the details. What we do know is that there is still a lot of debris hurtling about between the planets. We continue to be vulnerable to collisions with this debris and, if the earth should collide with a particularly large rock in space, we would have no advance warning and no protection.

THE BLACK HOLE INCIDENT

CHAPTER TEN

Peter Aller rounded the corner with the desperate urgency of a hunted man. It seemed as if he had been running forever, and nothing looked familiar any more. There were houses as far as he could see down the endless, empty street before him. If only he could find *someone* who would listen to him!

He raced up a walk into a yard and took the porch stairs three at a time, then pounded on the door. A heavy-set woman in a flowered dress answered it hours later.

"Can't you ring the doorbell? What are you banging for?"

"Ma'am, get out of the house! The earth will be hit and it's going to be awful! Get everybody out of there right away, there's no time . . . ."

"Oh, a nut, eh? Look, Mister, you'd better go take your story somewhere else. I got no time for nuts!"

"*Please*, Ma'am. . . ."

But she had already slammed the door in his face. Peter leaped from the porch and started to run again. Several times he saw small groups of people and tried to talk to them, but no one paid any attention. Was he invisible? Couldn't they hear his voice? How could he explain a black hole, rushing straight toward the earth, about to swallow them all up?

A huge, brick house loomed up; it looked like buildings on his college campus. Once again he ran up and pounded on the door. A man in a white coat answered. As Peter began to blurt out his message he noticed something familiar about the eyes, the beard. My God, he thought, it's old Margoulis, my biochem prof.

"Dr. Margoulis! Thank God I've found you. It's me, Peter Aller, and you've got to listen. There's a black hole. . . ."

"I know, Peter. It's all right. Just come inside and sit down. Everything will be all right."

Peter followed the professor like a tired puppy and sank into the chair indicated. "You have no idea. No one would listen! But it's going to smash right into us!"

The professor stood looking at Peter strangely. Slowly Peter realized he was in a very large room with a big, elaborate computer complex arranged against two walls. Wow, he thought, Margoulis has a whole setup, right here in his house! His head began to swim as he watched hypnotic lights flashing on the panels. The professor was hunched over one of them now, pushing various keys.

"Are you working on the black hole, too?" Peter asked, incredulous.

"You'll see, Peter. Just relax for a minute, and you'll see."

Suddenly a screen lit up and a strange face appeared on it. The eyes were slit and gleaming, the head bald and domed high, an alien face that looked intently at Peter.

A voice came from the screen: "I am Titius Bode. I

see you have summoned Peter Aller before me. Your work is well done, professor."

Peter froze in his chair and looked questioningly toward Margoulis, but he seemed to have left the room. Only the screen figure was present, and now, as Peter watched, the head grew up over the top of the screen and a life-sized figure followed it, jumping down onto the floor just as its feet emerged. Unable to move, Peter stared helplessly.

"From the planet Asta I come," the figure said in a menacing, gravelly voice. "You are the one who directed the black hole at my planet, and now you draw it here to earth. But you will not wipe out another planet, I tell you, for I will see that you alone are destroyed, Peter Aller. This time the black hole will be directed at you and only at you."

Peter tried desperately to open his mouth and speak, but it was as if he were made of poured concrete.

"In exactly one minute, the black hole will emerge from this computer. It has been programmed by the professor and myself. It will harm nothing until it reaches you, and then you will be swallowed up instantly." The words of the figure that called itself Titius Bode were followed by chilling laughter.

Peter struggled in vain to wrench himself from the chair, to speak. Waves of terror were sweeping through his rigid body when a loud jangling bell began to ring. Peter steeled himself against the terrible pressure he expected, but the bell just kept ringing and ringing.

Finally one of Peter's eyelids flickered, and the room took on a different appearance. He seemed to be lying on the floor now, and the jangling bell was somewhere overhead. "Oh, God," he said aloud as he recognized the sound of his telephone and fumbled up toward the edge of his desk.

"Peter Aller," he managed to croak into it.

"Your program's run, Dr. Aller," said a pleasant female voice at the other end.

"Oh, thank you. Thank you, I'll be right down."

Peter slid the receiver back into its cradle over the edge of the desk and fell back onto the old, worn cushion he had taken from his armchair last night when he decided to catch a few winks while the program ran. He lay on the floor a few more minutes, replaying the dream and getting his bearings. Through the single tall window of his cubbyhole of an office he could see that it was dawn. His desk held its familiar clutter, papers hanging precariously over the edge. Besides the desk and armchair, there was only space for the high tiers of shelving that held every book on astronomy and other sciences he had ever owned. Plus his journals and all the papers he had delivered at seminars and meetings. A thin imitation oriental rug lay under him, and a sprawling plant in need of water hung in the otherwise bare window. The quarters of a very junior astronomer here at the institute, he reflected, but comfortable enough.

The program! Pulling himself to his feet, he dug his fists in his eyes and ran his fingers through his hair, then loped out into the corridor. Checking himself, he decided to walk. There had been enough running in that dream! Besides, he was in no hurry to be proven wrong by the results of his program.

Titius Bode! Suddenly he laughed aloud as the meaning of the name came to him. The 100-year-old Titius Bode law predicted that a planet should exist somewhere between the orbits of Mars and Jupiter, where, in fact, only the asteroids had been found, the asteroids whose study had led Peter to his black hole prediction. So his murky subconscious had not only invented an ancient planet but had given it a survivor who now blamed Peter for a black hole cataclysm!

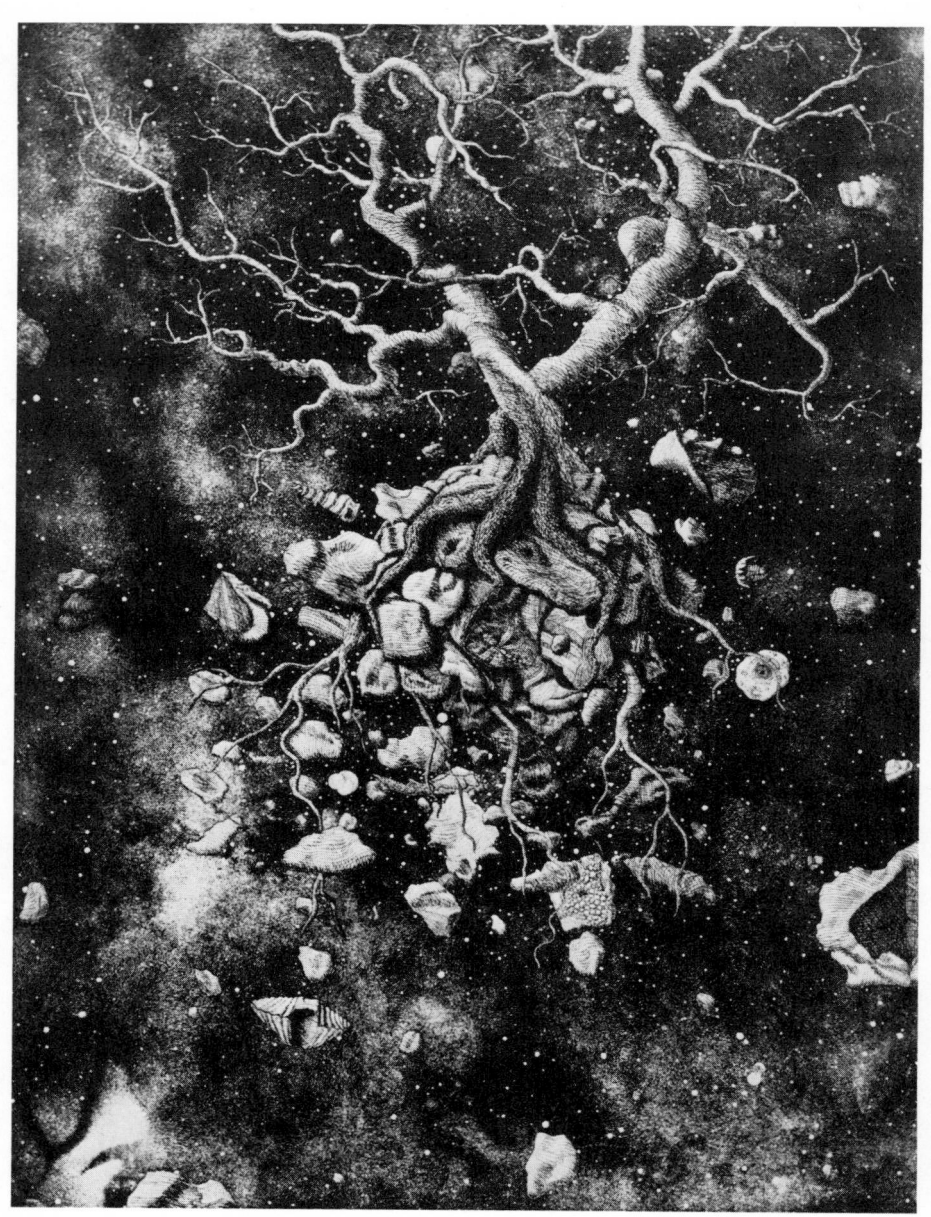

Asteroid. Etching.

Of course, the life-on-Asta idea had come from the discovery that meteorites from the asteroid belt contained some of the very basic molecules necessary to life on earth. He still recalled Margoulis talking about that when he was a junior in college, and that explained why Margoulis was in his dream.

No one knew why amino acids should occur in asteroid rocks and nowhere else except on earth, not on Mars, not on the moon. That was one of the big puzzles, but there was an even more basic mystery in the asteroid belt itself. What were thousands of pieces of rock, ranging in size from pebbles to 800-kilometer diameter planetoids, doing in orbit out there? At only one-thousandth of the earth's mass all together, there was hardly enough to account for a broken-up planet. But if they hadn't been smashed apart, why had they never joined to form a larger body? None of these questions had ever been answered.

Yet, after all the years Peter had pondered them, the idea of a black hole that could have eaten up most of a planet on impact, spewing the remaining fragments into space, had never dawned on him until last night. He'd been out for a beer with Jake Evans, then stopped by at his office before going home. For some reason he'd been thinking about invisible masses, and a chill went up his spine as the puzzle suddenly seemed to fall into place. He remembered how his mind had raced as he thought of the catastrophic shock waves that would be generated in a solid planet as a black hole passed through it, tearing it apart even as it was sucked into the powerful vortex of the invader. Whatever fragments weren't sucked in would have been thrown out into space—some going into highly elliptical orbits around the sun and occasionally crashing down onto other planets and moons. Some of those might become moons themselves, of planets like Jupiter, whose outer moons didn't seem to match the planet they orbited. Others would stay in orbits not very different from that of their original planet.

A small black hole in a similar orbit would not be detected by studying the orbits of other planets—it would be too far from them—but it might affect the orbits of the asteroids themselves. And having swallowed the additional mass of Planet X, the black hole would grow even more powerful. What if it still lurked out there? Some of the asteroids' orbits crossed those of the earth in their annual trek around the sun. Eros, one of the larger ones, had come dangerously close to earth in 1930. Here were all the institute astronomers, talking about supernovae and galactic dust clouds as dangers to earth when all the time an invisible monster might be plowing around in our own solar system! Those were the thoughts going through Peter's mind as he had taken the data on asteroid orbits and added an unknown object, of unknown position, orbit, and mass. He would ask the computer to figure out its characteristics. If his calculations worked, he'd know why the asteroids moved the way they did, among other things. . . .

At 3:00 A.M. he had finally been satisfied that his program would run, and he took the data tape from a shelf and carried it down to the computer center. As Peter entered, the night operator looked startled. It had been a slow night, and the operator had been deeply engrossed in his magazine.

Peter watched as the tape was mounted and the card reader rapidly read his deck. His heart sank as he heard the line-printer clatter within its sound-deadening cover; it meant that the program had failed. Tearing off the print-out he'd scanned it and immediately realized his error. In his haste to submit the program one of the cards had been inserted out of order. The error was quickly corrected and this time there was only a brief clatter as the printer told him that the job was begun.

"How long do you expect this job to take?" the operator had asked.

"I don't really know. It could be an hour or so. It depends on how much time it gets from the other run you have in. Call me when it's done, will you?" The operator had acknowledged affirmatively as Peter left the center.

Back in his office he had stood and gazed at his scribblings, still spread all over his desk. A black hole, he thought. Wow! He hardly believed it was true. Reaching for the cushion of his old armchair and tossing it on the floor, he had stretched out and adjusted the cushion under his head as he lay staring at the ceiling. What if it really is there? What if its orbit is also eccentric and crosses earth's orbit? That would be a scene! As his eyes closed, he had wondered about the reaction he would get from his colleagues if he was right. Or from everyone else, for that matter.

He was still a little dazed from his nap when he entered the computer center again. The night shift was over and the new operator, Mary, was already at her job. She indicated which pile of output was his. He gasped at the size of it; he had told the computer to print out results at several critical steps along the way, but he hadn't expected quite this much. He leafed through the pile of paper not daring to go straight to the last page. But it was obvious: The orbits of the major asteroids could best be explained by assuming that there was another asteroid of at least a hundred times the mass of the largest one already known! The final page gave him the confirmation he needed. Clearly there was a large mass out there, but something that big should have been discovered a hundred years ago. Since it wasn't, it had to be invisible, and the only object that could be invisibly wandering through the asteroid belt had to be a black hole.

He sat down and stared at the operator. She was pretty and seemed to be a very pleasant woman, though he knew her only through occasional contacts on this morning shift. He usually arrived at work after she had gone. Several times he'd seen her trying to cheer up scientists who were struggling with their programs, especially when it was obvious they had bombed for the tenth time.

"Why are you up so late, or is it early?" she inquired lightly. "It'll probably help to get the thing working if you get some sleep."

"That's not the problem. The problem is that it *did* work. Besides, there's no way I can get any more sleep until I try one more thing. God, this is the most incredible thing that anyone has ever discovered!"

"I've heard that before," she grinned. "You crazies are all the same."

"No, really! Looks like there's a black hole in orbit around the sun. I don't know yet where it's moving, but it might cross earth's orbit. I've got to run another program right away and see if I can predict its position before I take off."

"There are a couple of other runs in before yours," Mary responded. "You'll have to wait."

"Oh God, no. I can't wait. Look, I'll go get it, and it won't take ten minutes to run it. You can sneak it in. It's really important."

"Well, hurry."

He ran back up to his office and soon returned with another deck of cards. In the keypunch room he typed a half dozen additional cards, using some of the results from the previous run. He handed her the deck of cards, saying, "Okay, let's go. The future of the earth might depend on it."

"You *are* crazy," she muttered with a smile, taking the cards.

His heart began to pump a little harder as his imagination began to play with the possibilities again. If the black hole that had torn planet X apart was really in an orbit that intersected earth's orbit, then there *was* a small chance that it could come close to the collision of his nightmare. The idea was absurd. He refused to speculate further. But his mind was now warmed to the idea. If the black hole crossed earth's path, it might well cross Mars' path, too. Perhaps that was what had caused Mars to change its tilt and freeze up. The Viking photographs revealed that Mars had once been very active; water had flowed over its surface and river beds were still evident everywhere. Some time in the past things had been very different on Mars, but then it had changed dramatically. . . .

"It's done!" Her voice shook him out of his reverie.

"Pass me the output, please. I don't think I can even stand to look at it."

Mary came over to him with a smile. "Looks good," she said. "It didn't bomb." She handed him his sheets of paper and waited as he spread them out on the table. The hum of the power supplies and fans seemed louder as he stared at the data in utter amazement. There was a clear periodicity in the orbit of the black hole. It crossed earth's orbit regularly every few thousand years. His data weren't good enough yet to predict the next crossing, but it appeared to be in several years' time and unusually close to where the earth would *be* at that time.

"Well? Is it the big breakthrough?" Her voice was cheerful. This morning she sensed that there was something special about his earnestness.

"It's too incredible for words," Peter said, scraping the chair back from the table. "I have to check exactly where the earth will be."

"What are you talking about?"

He rapidly told her and left her standing open-mouthed as he returned to his office a third time to get the number of the tape that contained all the planetary orbit data. He would run the program once more and find out whether this black hole had ever come close to the earth itself or to Mars.

The results of his last computer run confirmed his worst suspicions. The orbit of the black hole was locked into those of both earth and Mars, so that it periodically came close to both planets. Apparently it would never actually hit them; that fate had befallen Planet X instead, a billion years before.

His colleagues began to arrive for work around 8:30 A.M., just as his final numbers were being punched out. Jake was irritated because his long program hadn't been completed during the night. Peter ducked out, but as the door closed behind him, he heard Mary say something about "earthshattering importance."

Two days later the topic was entirely out of his hands. It had taken a day to convince the director and his colleagues with a hasty seminar on what he had done. He had also calculated the mass of the black hole more accurately. Several local astronomers had photographed the specified location in the sky overnight and had found nothing. There should have been a body of mass 20 times that of the moon up there, but if it was there, it was invisible.

Confirmation for his theory came when NASA scientists, who had tracked several spacecraft through the asteroid belt, confirmed that they had already theorized the presence of some unknown mass in the belt to explain the path of the Pioneer II and Voyager spacecrafts. Their observations fitted the data he supplied. There was indeed a black hole in orbit around the sun.

The next closest approach of the black hole to earth would be in only a couple of years. The black hole would pass about two million miles from earth, eight times farther away than the moon. However, since its mass was 20 times the moon's mass, it was clear that the tidal effect on earth would be tremendous even if it was two million miles away. The physical size of this black hole was no more than that of a single atom, yet it contained as much matter as 20 earth moons!

Calculations showed that the tides it would produce would be two or three times normal. This meant disaster. Flooding would be rampant along all ocean fronts and the problem would be heightened if it were spring tide at the same time. If the influence of the distant black hole were to increase slowly the tides would be serious enough, but it was possible that complex interactions between continental masses and the moving water could trigger tidal waves that would be even more devastating. It was also realized that the accuracy with which the black hole's mass was known was not much better than 50 percent, so the tides could be much worse or, hopefully, somewhat less than expected.

Other astronomers calculated that a change in the tilt of Mars' axis a billion years before may well have been produced by this tiny but rampaging black hole. Some began to speculate that at its last closest approach to earth, some 12,000 years ago, it may have produced the famous floods referred to in so many mythologies, including the Noah myth. In those times, since no one knew the flooding was imminent, towns and villages must have suffered incredible destruction along the shores of the inhabited world.

Since the scientific community took the impending visit of the black hole seriously, national governments and the United Nations began massive evacuation plans. Virtually all the major cities along the coastlines of the world would suffer massive flooding, repeated day after day as the tides came in. This would necessitate evacuation of all ports for at least three months. Only then could the people return again, provided there was anything to return to. Millions of individuals would suffer inconvenience, but the physical destruction of the ports would be catastrophic. They would suffer so much damage — especially if tidal waves were generated — that they would effectively be destroyed in one fell swoop. In the absence of tidal waves many ships could survive by being out at sea, but if there *were* tidal waves even they would suffer the risk of total destruction.

Peter Aller, in the meantime, had achieved notoriety. He was besieged by reporters and cameramen eager to interview the discoverer of this "earth-shattering event." One evening when the black hole was still more than 10 million miles from earth, he tried to catch a couple of quiet hours by watching a late-night movie on TV. Suddenly a shot rang through his living-room window.

Peter was hospitalized for several weeks, and after that he was glad to find he was no longer news. His would-be assassin had turned himself in and confessed, raving about Peter's sins being visited upon the earth. But most public attention had turned to preparations for the great flood.

BLACK HOLES
CHAPTER ELEVEN

Among the most bizarre phenomena in the universe are black holes. They swallow everything that comes within their deadly range. They are unimaginably dense. Nothing, not even light, can escape the pull of their gravity. We cannot even see black holes; we can only deduce their existence.

A black hole is produced when an enormous amount of matter is concentrated into a tiny volume of space. When an object shrinks so much that the pull of its own gravity makes it "swallow itself," it becomes a black hole.

The earth could, in theory, become a black hole provided it was compressed to a ball only a half an inch across, a concept that sounds absurd and staggers the imagination. There is no reason to believe this would ever happen to our planet, but astronomers do believe that black holes exist.

One way a black hole is formed is when the interior of a star collapses so violently and the infalling matter crashes together with such tremendous force that its constituent particles blend. If this happened to our sun (a most unlikely event), it would collapse to a ball only 6.5 kilometers across.

To understand what happens to make a black hole disappear you should bear in mind that gravity pulls on all things, and that includes light. In a sense, light has mass. When light passes close to a large object such as our sun, it bends off its otherwise straight path through space. The larger the mass of the star, the greater is the bending that light beams will undergo. If the mass is extremely large and is concentrated in a very small volume, then light rays passing close by can even be forced to begin spiralling into the object. It is then possible to imagine that if a large enough mass is enclosed in a small enough volume, the light could be bent so much that it couldn't get past at all and would instead be sucked in. When that happens the object has reached the proportions of a black hole. Any light that tries to leave it for outer space simply cannot overcome the staggering pull of gravity. Since light cannot escape from a black hole, we are never able to see one directly. Black holes are invisible.

Each black hole has a distinct boundary. Inside this boundary you are effectively gone from our universe. Just outside it, if you try very hard, you might still escape its deadly grasp. This boundary is called the event horizon. You cannot see through it, and anything that falls beyond it disappears forever.

One way to picture what happens near black holes is to fly a spaceship into one, an unlikely mission indeed, but one that can be theoretically studied using modern physics.

The first important point is that the pull of gravity becomes so enormous at the event horizon that as you get close to it, it affects your body in a very dramatic way. Imagine going down feet first. Because your feet are closer to the black hole than your head is, your feet would feel a stronger pull than your head. When you are about 3200 kilometers from the black hole (if it contains the equivalent in matter of about 10 of our suns) you find that the pull between your head and feet, which is becoming greater and greater all the while, is so great that your body is no longer able to counteract this pull with the power of your muscles. All the while you are falling faster and faster into the hole. Because your feet are being pulled so much faster than your head, you are quickly stretched out. At the same time the gravitational forces operating on your shoulders tend to bring them closer together. Of course, the human body isn't meant for this kind of torture, so you won't survive very long past the 3200-kilometer mark. As you hit the event horizon your body is very long and very thin, like a piece of spaghetti. Since spaghetti is not too aware of itself, perhaps we had better take a more distant view of this experiment. Let's watch what happens when a spaceship with someone else aboard falls down and down into the black hole.

There is a side effect that results from the enormous gravity of a black hole. Time no longer flows the way we are used to. If we watch the spaceship falling down, we find that it seems to move slower and slower, and as it gets to the event horizon it seems to slow down to a stop. In fact we would never see it disappear. At the same time, the light emitted from a beacon on the ship would become steadily redder and redder. This effect is expected when objects such as distant galaxies move rapidly away from us. This redshift is a well-understood consequence of the stretching of light emitted by a receding object. However, a redshift can also be produced when light struggles to escape the pull of gravity from a very massive object. Thus the falling spaceship appears to move slower and slower even as the light reaching up appears redder and redder. At some point the light is stretched so much that it becomes infrared radiation (a form of radiation similar to light, but which we feel as heat).

When the spaceship is even closer to the event horizon, the emitted light gets stretched out so far that it is similar in wavelength to typical radio waves. In a sense, then, the spaceship becomes invisible long before it disappears over the event horizon.

On board the spaceship things are quite different. To any astronaut on board everything would seem normal (provided the astronaut could avoid turning into a spaghetti-like mess). There is none of this apparent slowing down, and the craft will seem to fall unimpeded beyond the event horizon and into the black hole. This strange contradiction is a consequence of the way black holes affect space and time around them. Scientists best understand this by using Einstein's theories of relativity.

More peculiar still, all things lose their identity in a simple black hole. Matter is no longer in any form we could recognize. An airplane, car, or human being all become indistinguishable from each other; only their mass is preserved. Also, nothing can now escape again. Matter continues to fall toward the center, and there one finds a very weird phenomenon. There, all the mass is piled up inside a space that has no volume! This is called a *singularity*. The laws of physics and time as we know them do not exist at the singularity. In fact, time as we know it doesn't exist inside the black hole at all, and if you were inside the black hole you could only move in one direction: inexorably toward the center, forever. Where does the matter actually go that piles up at the singularity? Some far-fetched explanations have been considered.

It is possible that a black hole in our universe connects with another universe, and the matter could continue right on through into that other universe. Beings in that universe would see a pouring out of energy from some location, and they would call that a "white hole." The two would be joined by a "wormhole." It is possible that such wormholes can also lead from their black holes to our universe's white holes. In a wild imagining you could travel in and out of universes by entering the correct black holes. In theory it seems as if you could even return to our universe before you left, since time in black holes appears to be no barrier. This possibility has caused considerable concern to theoreticians. It would also lead to absurdities, as can be easily illustrated.

Imagine that you climb aboard a spacecraft and fly into a local black hole. Somehow you survive and quickly find another black hole in that other universe and re-emerge in our universe a few minutes before you originally boarded the spacecraft. We can perform this thought experiment since we believe we understand the laws that govern black holes. It is also cheap! In any case, you will now see yourself getting

Anatolian Gateway. Oil on canvas. Time and space are transcended as we engage in another reality.

ready to board the spacecraft (or you will see yourself approach as you get ready to board). Then you can take your double along, in which case there will suddenly be two likenesses walking up as you get ready to leave, because both of them must have gone on the round trip, etc., etc. This is clearly absurd. Alternatively, you can say to your double just before you leave, that it is a hellish trip and you'd rather not do it, so you don't, but in that case the double wouldn't be there to tell you it was hellish, so you go, but then you don't go, etc., etc., which is also absurd. In fact, time travel is always absurd because of these contradictions. There must be some way in which the universe and black holes behave to prevent such a trip into a black hole, even in a thought experiment. What is the catch?

The way around this apparent dilemma was dreamed up by physicist Stephen Hawking who is renowned for his work on black holes. The explanation, which is quite esoteric, involves the production of particles by a black hole. In order to picture this we need to consider an interesting aspect of particle physics that physicists have been aware of for quite some time. First of all, matter as we know it consists of electrons, protons, and neutrons, the three so-called building blocks of atoms. In addition, we know of the existence of antimatter because it can be produced in laboratory experiments. Each of the three particles has a corresponding antimatter particle. The electron has its positron, the proton the antiproton, and the neutron the antineutron. The first two pairs differ in the electrical charges they carry. The neutron carries no charge; nor does its antiparticle. The electron carries negative electricity, while the positron is positively charged. If the electron and positron collide they cancel each other out and disappear in a flash of energy. The name for the radiation produced when antimatter annihilates matter is a *gamma ray*. The reverse process is also possible. If great beams of energy (gamma rays) collide, they can produce both a particle and an antiparticle. This is known as pair-production.

Another useful concept in physics imagines that the potential exists for pair production anywhere in space, provided you can zap enough energy into any one spot. Thus we can imagine that there is a potential pair of particles virtually anywhere in space, called a *virtual pair*. There is a virtual pair (an electron and a positron) at the tip of your nose, and if a large blast of gamma rays were to hit it you might see an electron and a positron speeding away as if created out of nothing. Such an experiment would be hard to perform, but is again allowed in thought.

What has this to do with black holes? The answer is that at the event horizon the gravitational forces are also so great that they can in fact generate pair production. A black hole therefore produces enormous numbers of matter particles and antimatter particles. The outcome of this is that the door to the black hole is closed, because you cannot survive the flight through these particles to reach another universe no matter where it leads.

We have been discussing a simple black hole, one sitting quietly in the universe, minding its own business. Most black holes, however, especially if they formed from stars, are expected to be rotating. A rotating black hole is a much safer place to visit! Around a rotating black hole there is a region of space just outside the event horizon in which you can actually dally for a while and still hope to escape to outer space again. The black hole will drag you around with it, but the fascinating thing is that if you were to fly into this region (known as the ergosphere) you could emerge with more energy than when you entered. A rotating black hole turns out to be an inexhaustible power supply. For example, if you dropped a large boulder into the ergosphere and if it were to split while inside, one half could fall into the black hole but the other half would come hurtling out with more energy than it had when you threw it in. This concept led some physicists to a great, impractical idea for constructing a perfect ecologically safe, and very unlikely city. You construct a shell around a small rotating black hole and build your city on the shell. Then you construct a garbage disposal system with a series of hoppers carrying garbage down into the ergosphere. There the garbage is dropped into the black hole, and the hoppers emerge going faster than when they went in. You then hook the hoppers to an electrical generator that takes some of this added energy and converts it to electricity for use in the city. This would be an infinite energy source, thanks to the garbage you generate. Of course, you have to be careful not to feed the black hole too much, or it will grow and swallow the city!

This leads to another fanciful use of a black hole. Again imagine a sphere around it, but this time line the sphere with mirrors on the inside. Then cut a small hole in the sphere and shine a flashlight into it. The light will bounce around inside, and every time it passes through the ergosphere some will disappear into the black hole. But much of it will be amplified, that is, get brighter, as it bounces around inside. If the hole is still open you will find a strong beam of light shining out, even though you injected only a little light. If you were to plug the hole then you would make a bomb, because the light would go on and on bouncing about and gathering more and more energy until it is too much for the shell to hold and the whole thing explodes very violently.

These are very speculative thought experiments concerning black holes. They suggest that if we could catch one, and could handle it very carefully, we would have infinite sources of power.

But do black holes really exist? Astronomers believe they know of at least one, but it is possible that they are all over the universe, if certain properties of our universe are what scientists now believe. It is even possible there are small black holes in our solar system.

It may be possible for a black hole to be formed when the inside of a massive star collapses catastrophically at the end of its life. Unlike other stars, where the center implodes to become a neutron star and the rest of the star's mass bounces off in a supernova explosion, the black hole fate awaits only those stars that are much more massive than the sun (containing at least 10 times as much material). In those cases the initial implosion forces matter so close together that it becomes a black hole. The whole mass of the star disappears into the black hole and the star doesn't explode; it simply disappears.

Such a black hole can be observed provided another star is in orbit about it. Material can flow from the visible star and fall into the black hole. This matter does not crash straight down into the black hole, but spirals rapidly inward, and it can heat up to millions of degrees before it disappears forever. During this death spiral hot gases begin to emit strong x-rays that can be detected by x-ray telescopes in earth orbit.

Astronomers have found just such x-rays being emitted by something in the constellation of Cygnus. This source of x-rays is located at the position of a faint visible star, and careful observation shows that this star must have a very massive invisible object in orbit about it. Since the invisible companion is too massive to be a neutron star, it must be a black hole.

There may also be many black holes formed when single massive stars imploded. These would have no companion stars in orbit about them. Just how many of these exist in space is unknown, because these black holes will forever remain invisible. There is not enough material falling into them to make them shine by emitting x-rays.

Besides the known black hole in Cygnus (called Cygnus X-1) there may be black holes around left over from the birth of our universe. These are even more remarkable, for they can explode from time to time. If true, they may have interesting consequences for life in the universe.

When the universe was very young, only a few seconds old, things were pretty chaotic, and the forces exerted on any tiny fraction of space were incomprehensibly large, large enough to produce black holes. Imagine for a moment that you wanted to turn a brick into a black hole. It would have to be compressed to less than the size of an electron. Clearly there is no way we can conceive of ever doing this, but in the early seconds of the universe it is possible that black holes of all sizes were created. (These are referred to as primordial black holes.)

Small black holes turn out to have very different properties from large black holes. We mentioned before that the gravitational pull gets stronger and stronger the closer you get to the event horizon. For a black hole with 10 times the sun's mass you start to feel its influence thousands of kilometers away. For a tiny black hole of thousands of tons' mass, however, you feel its gravitational tug only when you are very close to it, just a fraction of a centimeter away. Then the gravitational tug suddenly becomes enormous as you approach ever closer. This tug is so great that space immediately around this little black hole contains a staggering amount of gravitational energy, so that pair production is easily triggered.

The material produced by pair production finds itself in a tiny space, and now we find that another strange property of matter on the smallest scale comes into play. The particles and antiparticles can "leak out" of the black hole, provided the hole is small enough. One can imagine that by the time the particles are created they have already wandered out of the black hole. This effect, known as tunneling, is observed in the laboratory in more realistic experiments using ordinary matter. If particles or antiparticles tunnel out of the black hole, we effectively observe the black hole to be emitting radiation. An object that is emitting radiation in this way can be described as having a temperature; it is effectively radiating heat in the form of particles. The black hole quite literally begins to evaporate away. It is even possible to calculate the temperature of a small black hole like this. A million-ton black hole, for example, is radiating energy as if its temperature was a quadrillion degrees! As the hole gets smaller and smaller it gets hotter and hotter and (in this example) after three years it would be all evaporated away. But then we find another amazing phenomenon coming into play. These tiny black holes are expected to explode violently just before they cease their existence.

As they evaporate they get smaller and smaller, making

The spiral galaxy NGC 2903, which contains 100 billion stars. How many of those stars have inhabited planets in orbit about them? How many black holes drift between those stars? No one knows. Hale Observatories photo.

it easier for the particles to tunnel their way out. They therefore evaporate faster and faster. In the final seconds, this evaporation is so fast that it is effectively an explosion. But this is no ordinary explosion. The black hole disintegrates with a blast equivalent to a billion-megaton bomb. Small black hole explosions of this kind must have been frequent in the early universe. Some may still occur today, and astronomers have launched searches for such events.

Of the primordial black holes, all the smallest ones soon exploded. As the universe aged only the originally large black holes persisted. Some of these may have evaporated down to a fairly small size by now, and others may be exploding somewhere in space.

A black hole that was originally about a billion tons in mass would last three billion years before it exploded, while a starlike black hole will last much longer than the age of the universe. The Cygnus black hole, for example, being several times the mass of the sun, is never expected to explode.

It is even conceivable that space immediately around us is filled with black holes of all sizes. Could some of them be in the solar system? Could one go off someday? No one knows. This is an aspect of black hole research that is highly speculative, and research in the next few years will no doubt tell us much more about these strange objects.

As a final thought about their importance, we mentioned before that the solar system might have been triggered by a nearby supernova that compressed an interstellar cloud of gas and dust, which then collapsed to form hundreds of stars, which then gave rise to supernovae to trigger the birth of the next generation of stars and so on. Perhaps it was black hole explosions, that triggered the first generations of stars, black holes formed at the big bang. Perhaps it was black holes that started the life-death cycle of stars that eventually led to the formation of our planet and our existence.

The interplay of forces in the universe is truly awesome. Our existence can well be related to the existence of virtually everything else in space and time.

RED GIANT
CHAPTER TWELVE

A single beam of light fell straight down on the
Supreme Initiage. He stood on the raised dais at the
front of the huge temple, his back to a thousand hooded
figures. With his arms outstretched and his feet wide
apart in the symbolic stance of a five-pointed star, he
displayed the white unicorn blazing against his black
robe. He turned slowly to face his select audience of a
thousand lesser Initiates.

"I, Makara, having taken the name of the penta-
gram and the symbol of the unicorn, herald the Age of
Capricorn. Transcending the ambition of the goat, the
unicorn represents world salvation through victory over
all tests. Integrating the five-pointed nature of man with
that of the Saturn-Father, the unicorn brings together
matter and spirit — the two poles of Mahat, the Universal
Intelligence. The forces of Ahrimanic darkness are dis-
integrated in the Light of Capricorn, bringing harmony
in place of self-serving ego.

"The Christ, born under the sign of Capricorn, was
slain once on the cross and again by the Aquarians,
who failed to understand the practical tests of a material

world and sought to reach spirit without recognizing the true nature of transcendence through integration. . . ."

Nyra had placed herself strategically close to one of the exit shafts and ducked into it the moment the ceremony was over. Flinging off her robe, she jammed it into the nearest open lock-tube and jumped into a waiting transporter before any of the others could catch up with her. As she was whisked down the shaft, she swept her hair smoothly around her head and pinned it with the clasp of a Class-I scientist. At the Omicron junction, she looked around again for any observers, then jumped an express transporter for the Center.

She knew it took exactly 10 minutes to cover the remaining 31 kilometers, so she settled back in the little bubble to catch her breath and think. She'd be less than a minute late for the special flow-check at 0200 and, if necessary, she could fabricate an excuse easily enough. It was hectic managing a dual life, but so far, so good.

When the bubble hissed to a stop, she leaped out and ran to the vertical shaft marked by the red light. Inserting her code key and stepping inside, she egged it on mentally until it reached bottom, where she grabbed another transporter and headed for the Sensor Complex. Nyra had seen tapes of the ancient little mine cars that once trundled down these deep shafts, and now she praised the unicorn for her efficient vehicle. Even before she hit the halt disk, the throb of the pumps reached her.

Adjusting her belt in a gesture of composure, she inserted the key in a massive door and walked through briskly. On the other side were huge tanks of fluid connected by a complex array of conduits and control instruments. In a far corner of the cavernous room, more than a thousand meters wide, she could see the gleaming panel of lights on the main computer. Her colleagues, Tai and Vintak, were already in conference with the computer operator, and both flashed looks of reproach as she joined them.

The operator, also a Class I scientist, spoke intensely as he hit a panel switch that instantly deposited a sheet of filmy material on his desk.

"Here's the summary. As you can see, the neutrino count has been increasing steadily for two weeks. It's quite obviously not due to system fault. Look, here, and here—everything checks out. The detectors in the other units confirm the count."

"So we saw," Vintak acknowledged. "There really seems to be no chance of error, does there?"

"Quite." The operator hit a number of disk switches in rapid succession and stood up. "Might as well get up there. Nothing's wrong down here. I don't know why those people doubt the Sunsensor, but I suppose, under the circumstances, we have to be absolutely sure."

"That's for cert," Tai agreed with a typically humorless grimace, "with the future of the planet at stake! "

The four scientists boarded a runabout and drove back to the vertical shaft. The sound-lift rapidly raised them to ground level, where they took another runabout through the corridors to the Main Unit. Nyra pushed back all thoughts of the ceremony and now scanned her memory patterns to see if anything had been overlooked. The message was clear from the few words she had heard and from the data; she found no loopholes. Later she would have to play out the implications for the Followers, but right now she was among those who knew nothing of her true allegiance.

They passed the plaque that commemorated the original building on this site. A very crude neutrino detector had been built into an abandoned gold mine here in the old "Dakotas" back in the twentieth century. The first Sunsensor Center replaced it in the next century, when the great breakthrough in detectors came, but the Dark Ages began not long after that and the first Center was destroyed. Now, two and a half thousand years after that original detector, the giant Sunsensor Complex neared a thousand years of peaceful operation. It had become a gigantic underground network unimagined by those twentieth century scientists, Nyra thought. She herself marveled at its size and complexity.

They left the runabout in front of the main organization building and strode into the Chief's office quarters.

"Sorry we had to resort to such a crude way of checking." The words came from the man they called Chief, whose tall angular frame stooped over the control desk. "I suppose it's fitting that we should introduce a human touch, even at this late stage. Everything as we thought?" The four blueclad figures nodded in unison as they seated themselves at terminals along the conference desk. The Chief left the control desk, touched the door lock, and joined them.

"The President has already been alerted and is standing by. If you're all ready, I'll switch her right in. She may have some questions that you should answer directly." The Chief touched his video screen lightly and six images of the expectant World President appeared around the great oval desk.

"President Karman, we have done all we can and our report is final. The neutrino count is increasing steadily. We estimate that. . . ."

"Just a moment." The President's voice was somber. "I have taken this opportunity to call the entire World Council to their control units and we also have contact with Mars, Amalthea, and Uranus. Please give us your report and your recommendation, as well as an estimate of the margin of error. I've heard some very wild figures quoted." She shifted in her seat as if to prepare herself for the decisions she was about to make. Her body seemed tense as she leaned forward and gave them her attention. "Proceed."

The Chief quickly repeated his first words. "It's all checked out. A complete diagnosis was run and, as a final resort, we had several scientists go down there to make sure nothing had been overlooked. The count is already at 2750 SNUs and the increase seems to be steady at 10 percent per month. Based on the measurements made back in 2196 by Enterprise III on her automated journey into the Gould's Belt star system, such an increase preceded the red giant phase of the star Wolf 684. Shortly thereafter the neutrino detectors at Sunsensor I also observed an increase, but only over the last few months before the expansion of that star. It's fortunate for us that Enterprise III happened to be only five light years from Wolf 684 and could witness the formation of that red giant, so we have its data today. No doubt you are familiar with all that." He paused.

"Now, based on our predictions made yesterday, the sun is clearly entering red giant phase one. We estimate no error in this prediction at this time. We are clearly detecting neutrinos emitted by the conversion of carbon to oxygen. Our best computer models suggest that this phase may last two or three years at most."

The President interrupted before the Chief could continue. "Doesn't a red giant develop much slower than that?" She looked directly at the video unit as if looking for a way out.

The Chief answered smoothly from his vast fund of knowledge. "The estimates have been revised repeatedly over a very long time. Before the Dark Ages, scientists were already aware that most stars end as red giants, but they naively believed the sun wouldn't reach

Self-Extinction of Homo Sapiens. Oil on canvas.

that stage for billions of years. You will recall that the invention of the Sunsensor at the end of the Second Dark Age dramatically improved our knowledge of astronomy through neutrino detection. The source of these remarkable particles could actually be pinpointed with the giant neutrino telescope, and during the last thousand years we've discovered about 200 different sources. Still, the most dramatic one was Wolf 684.

"It's evident now that we really are able to predict the life cycle of stars with complete accuracy. Do you remember how the announcement that the sun was not expected to live much longer than another 5000 to 5100 years was scoffed at by everyone except those scientists actually working on the data and the theory? When the crystal computer came into service, it only took a few years to predict the exact fate of the sun. Our new data simply confirm the prediction."

The Chief paused and looked around the desk as if searching for help. Surely he wasn't expected to ramble on about work done at the Sunsensor, and all his viewers knew the work of the New Religion, whose adherents had been in power since the sweeping world elections almost a hundred years before. Fanatics, the Chief thought ruefully, who condemned the Scientists, yet allowed a technically-oriented woman to become President and used the power of science to their own ends.

The President broke in again, impatiently, as if she, too, were considering the political implications of this report. "Go on, is that all you have to say?"

"President Karman, it is undeniably true," the Chief continued, "that the sun is proceeding on its red giant track." He swallowed nervously as he came to the heart of the report. "We expect the consequences to be observable in the visible part of the spectrum sometime next year. Its heat output will rise rapidly for about three years, with an expansion phase lasting about two decades. This will swallow Mercury and reach nearly out to Venus. Ecological constraints on earth are such that irreversible change on our planet will proceed according to our predictions, and the earth will be unlivable within two years after that. We hoped that this increase in neutrinos would prove to be cyclical, with a return to normalcy in a year or so, but the count now indicates that the interior of the sun is in a new energy generation state, and nothing can reverse it. Only when it is a fully mature red giant can we expect it to erase itself. . . ."

As the Chief went on, Nyra almost pitied him. It was bizarre to see him in the position of having to predict the end of the world. How many people had done that in the millennia since civilization began? Not one of them has been a leading scientist, backed by the most advanced technology and the most precise data. Yet even now, with the intense polarization between the Scientists and the Followers, there would be many that believed no prophet of any kind. Strange how she, who gave allegiance to both camps, was forced into the role of subversive. There was a long line of believers before her who had tried in vain to reconcile science with religion. It seemed so reasonable, so obvious, and yet was so difficult in practice. As soon as this conference was over she would go to Makara.

"What about the planetary nebula phase?" the President was asking.

Well briefed, thought the Chief, as he prepared to respond. The enormity of his role was getting to him, and he felt a mounting pressure in his head as he pulled himself together. "That phase will be expected just 1050 years from now, with the first shock reaching earth about two days after the upper shell starts its expansion." He paused again, acutely aware of his own vulnerability, then continued, "I recommend *immediate*

implementation of Evacuation Plan Uranus." Swallow-
ing hard, he sat back.

The President showed no reaction to his drastic
suggestion. "Thank you," she responded courteously. "I
trust all the data are on the common link? If there is
nothing more. . . ?"

She was clearly anxious to switch modes for her
next conference. The Chief shook his head silently at
her and she blinked off screen. A chill filled him as he
stared at the blank space where she had been. It was all
entirely beyond his control now; he didn't even merit a
place in the Evacuation Plan. He looked at his colleagues. None of them did either—or did they? He
would never know. It was the best-kept secret the
planet had ever seen.

Without speaking, they all stood up and left the
office together. Outside, they stared at the blue sky for a
while until an empty transporter slid to a halt.

"I'll walk. Till later." The Chief lifted his arm in
farewell and stepped quickly over the rail into the field
that separated the Center complex from the city.

"Must have been rough," Tai commented. "Well,
let's go."

"I'm not coming either," Nyra said steadily. "I
want to go over to Epsilon before I leave for the day.
See you tomorrow."

The Chief ambled about in the huge field, his mind
numbed and his head pounding. He could only feel the
transitory nature of his—and all—life. He had grown up
expecting to live a good 150 years, and he was only 95
years old now! The prime of life. It seemed such a pity. But
the universe knew no pity. He sensed the sun's warmth
on his back and felt another pang of despair. Everything
in the universe had a beginning and an end; nothing
lived forever. Not people, not stars. He wondered how
many civilizations on other planets had suffered a sim-

ilar fate. No signals had ever been picked up. Perhaps this was why — few survived long enough to communicate with others, and the time distance was so great. Stars simply didn't grow old enough. What a pity, what a waste. Or was it? His mind refused to order either his thoughts or his body, and he lay down on the grass exhausted.

It was so green. Intense green grass, and intense blue sky. It seemed so peaceful, and yet the earth had seen lots of cosmic violence before this. It had even survived neighboring star explosions, bathing the planet in deadly mutating rays, changing its atmosphere. For billions of years. And now. . . .

Nyra retraced her route to the Temple, hoping she would still find Makara there. In the bubble to Omicron she reflected on the silence that was imposed on all Class I scientists there today, in person and by video. Some means would be taken to distract anyone with access to the data, she knew. The President would decide what the public should know.

Makara was in his Temple quarters and granted Nyra immediate audience. It was not without design that the Temple was so close to the Center. She briefed him quickly, having kept him aware of the events leading up to today. When she finished, somewhat out of breath, he smiled serenely.

"Be calm now, Nyra. It is not for us as it is for them. You Initiates have long been prepared, and even the rest of the Followers will find it easier than the others. The Scientists preach their doctrine of a universe filled with separate and finite objects; their message will bring despair, of that there can be no doubt. Many of the Followers will despair as well, but we must work together to bring as much understanding as we can. I want you to gather to yourself all the Scientists you judge in any way sympathetic, to translate our language into theirs, as I know you can. Perhaps we can still help them gain more perspective, so they will feel and show less panic, rail less against the inevitable. Then, when the people are told, their leadership will be stronger."

"But Makara, events are too close to spend time mincing words. And if I speak out now, they'll expel me from the Center."

"I didn't say mince words *or* speak out. Use your intelligence, your rational mind, to make the translation. Not to avoid being expelled, but to bring them understanding in *their* terms. The gap between us is caused more by persisting prejudice and language problems than by anything else. Of course they don't believe in nonphysical existence, but their physicists are getting very close to the reality of that.

"They mock our rituals because they still don't believe in direct access to knowledge through altered states of consciousness. You know the Dark Ages wiped out scientific efforts in that direction. We've talked about it so often. Because science did not renew those efforts when it regained its prestige, we are still seen as some kind of spiritual aberrants from the past. It's my own failure that I have been unable to make peace between our attackers and our own overzealous defenders. Perhaps now, when even the Scientists will feel the need to believe in something greater, you can find a way. And if they throw you out, it won't matter, as long as you tried. There will be much you can do to get our own space program ready in time."

Nyra openly revealed her profound love for Makara by the widening pupils of her deep green eyes. How could the scientists believe this man a fanatic, or even irrational? She could easily translate the Followers' language into that of science, and most of their beliefs, too. All but the survival of consciousness. She and Makara had always communicated easily. He respected her scientific knowledge and lifted her spirit immeasurably. She was not afraid of the coming cataclysm, and she longed desperately to help her colleagues face it. But could she?

"I'll try, Makara, you know I'll try." When she talked with him, she wanted to save the entire world. And she prayed secretly that he would not be one of the four Followers to be sent off in the little ship to Alpha Centauri for a new cycle of human life. It was, of course, too much to hope. But she found it hard to face her corporal end without him.

As Makara rose and came toward her, she felt her heart pounding. His farewell embraces always left her weak in the knees.

Meanwhile, the Chief was busy making hasty plans in his office at home. The woman who was his house companion, a professional botanist, had puzzled at his mood when he came in. He couldn't tell even her what was on his mind, so he passed off his distraught air as lightly as possible and popped a pill to clear his head. Now he was lucid again, as he finalized the details of data conversion. The master control would be programmed to make the neutrino counts appear constant. If the readings decreased, he and the Class I's would know the count had slowed its increase, but *only* he and they would know how to interpret the readings correctly. Good. It should work.

Leaning back, the Chief ran his fingers through the remaining hair over his ears and thought about the Evacuation Plan. Ten thousand people were indeed a select few to be saved from this lump of rock in a corner of the galaxy. They were going to have to live first on a hostile moon orbiting that giant gas cloud called Uranus. No one even wanted to estimate their chances on *that* migration outward into the solar system, much less the likelihood that a smaller number would set off from there to Alpha Centauri. It was a conscious struggle to survive, never before attempted by any living species. Who knows how many generations would be born and would die on a spaceship? Maybe a hundred people would have a slim chance of reaching Alpha Centauri, and he wished desperately that he could be among them.

At least it wasn't as crazy an idea as the one dreamed up by the Followers! The space drive they planned to use was certainly faster, maybe one-tenth the speed of light, but it couldn't move anything with a mass of over a few hundred tons. And he doubted they'd still be alive to reverse its acceleration when they reached Alpha Centauri, even if the ship did get there centuries ahead of the Scientists'. Their *Unicorn* was to carry two men and two women. What if it *did* work? The Chief harrumphed out loud. "Not possible," he said aloud. Where could they recruit enough Scientists to plan and build such a project? Why didn't they just fly off in their astral bodies, if they were so good at those things? One thing was sure, they would accelerate their project if news of the increased sun activity leaked out—but he was taking steps to make sure that that couldn't happen.

Four billion earth people and only 10,000 to get away even as far as Uranus. And he had to keep the secret or the plan was doomed. If he was so important, why wasn't he in on the plan? Torn between sulking and mustering up in the line of duty, the Chief went to bed.

As the months went by, the senior staff at the Sun-sensor Center went about their tasks outwardly calm, bearing the knowledge of their own fate and of the world's largely in silence. They were expected to work as though everything was going well. The only hope now lay in keeping the count and the plan secret until after the ships had departed for Uranus. Each member of the staff nurtured a private hope of making the list.

If word about the plan got out, there would be social disintegration throughout the world. Vintak and Tai, always very close, discussed this one day and came to the cynical conclusion that the social chaos could hardly match what nature had in store. It wasn't that they wanted to leak the information, but it was getting hard to pretend at home that they didn't know about the threat to their lives.

The Chief payed close attention to the news these days, and at night he watched the skies intently. Occasionally, he imagined he saw an unusually bright satellite pass overhead, and he wondered if it were really a Uranus-bound ship leaving orbit. The only hint that anything was happening came on the Channel 80 news. A reporter noted that unusual activity was taking place at the main earth launch center and questioned why so many people were traveling to the launch center. There was evidence that not all of them *left* the center, at least not by known routes. That same reporter was not seen on the screen again. According to the video news he died in a transporter accident.

The inevitable leak finally came in December, nearly a year after the Chief had reported to the President. Another enterprising reporter discovered that about 2000 persons who had booked on world cruises in old-fashioned steamships had never returned to their homes or other places of daily activity. Hysteria mounted for three days, while even the Chief was besieged with calls and demands to know what the sun was doing. Rumors flashed into prominence and took on an increasing semblance of truth, as the rumormongers guessed more and more accurately. The President on her island retreat, as well as those in regional authority, were swamped with reporters who demanded to know what was going on. Having foreseen this eventuality, the President had the Center release a story claiming there indeed had been a transient and dramatic increase in the neutrino count, but that the level was now constant again. Presumably, the Scientists would have to reconsider old theories and make some minor adjustments. This small change in the solar constant was not to be misinterpreted in any way as the threat of a red giant. The count change had basically been a false alarm.

Demands to see Sunsensor data were met openly, as the Chief and the four Class-I Scientists assumed that the doctoring of the program would go unnoticed to all but their eyes. None of the others suspected Nyra of having violated their pact. The way she was talking nowadays about universal life principles and organizing factors was considered an idiosyncratic response to the strain they all felt. Only once, Tai accused her of sounding almost like a Follower, and Nyra had to turn and walk way to restrain the reply she wanted to throw back.

Makara took her into his confidence as soon as he made the decision to tell the world about the increased sun activity. He had spent three days and nights in meditation when he sent for her by telepathy. He could not convey substantial messages to her that way, but he could call her to him quite effectively.

"It's been almost a year now, Nyra, and I've watched the public reaction closely — as well as the Scientists' cover-up. The lid can't be kept on indefinitely, and my inner guidance is to tell the world of this new transition in our solar system before people go mad with rumor and misguided terror. When I make the announcement, all the Initiates will be strategically placed for local gatherings. Each of the Followers will be asked to make personal peace with as many of the others as possible. When they understand the magnitude of the coming event, they will respond as I ask. But I would like to make my speech as credible as you can help me make it. Even if people can't believe in their personal survival, we can at least help them understand their connectedness. Understand that whoever goes on the ships takes the others' life force with them. Will you help me?

"I'm ready, Makara."

"Good. The power of the unicorn is to integrate matter and spirit realistically. We must use whatever means are at our disposal. Most of the Followers will begin concentrating on inner harmony when they hear, and by the time we launch our own ship, they will be in accord with the necessity of taking only four in body while the others accompany them in spirit."

"How are the plans? I've wanted to ask for so long."

"Since you and I have shared confidences so long, I will share this with you. The four have been chosen."

Nyra's heart leaped into her mouth as he went on to say he would not be among them. Oh, praise the unicorn, she exclaimed to herself, trying not to show her relief outwardly. He would be here with her when the sun grew to swallow them up. Their spirits could depart together.

"Did you hear me, Nyra? What's wrong? I said I've decided you are to be one of the four."

"Me?" She was dazed. Two such strong emotions in rapid succession. "But, but I never thought. I mean, I never wanted. . . ."

"I want you to go, Nyra. You are young and intelligent. You combine the best of the Followers and the Scientists. You understand more about the universe than any of us. I never even stopped to think about it.

You were the first one I chose, and I thought you knew it all along. If you have any arguments, don't state them. Just go home and meditate; then you will understand."

The way he was looking at her, she could not begin to oppose him. He had always treated her as an equal, and she had always marveled at that, for it was not his way with others. He had the natural command of a Capricorn leader, and now, when it came right down to something so big; he had it with her, too. She simply nodded, felt his embrace more electrically than ever, and left.

Makara's speech was tremendously inspiring. He had no trouble gaining as much time as he wanted on worldwide video, and the people listened. They also believed. Someone was finally giving a reason for the unrest. Even the President was spellbound. In a way, she thought, she had to hand it to him. Without the cloak of religion, the news would have been unbearable and *someone* had to tell it before much longer. Never in the history of the planet, she was sure, had everyone's attention been focused on one figure for so many minutes. She herself perceived an almost psychic contact over the whole planet as she heard Makara speak. If only that kind of leadership were possible without religion. What *was* it about people, she wondered.

Even Nyra did not know how Makara had determined that 6000 people had already left for Uranus, but undoubtedly he had other sources besides herself. It seemed that two spaceships were even now in orbit preparing for the journey.

When the broadcast was over, the chaos expected by most Scientists did not materialize. Perhaps, they thought, humans really do not riot when they know for certain their death is imminent. What could anyone do?

Perhaps, some thought, a little religion isn't so bad when something like this comes to pass. Science could offer knowledge, but comfort was another matter. The violence of the universe was hard to reconcile with the beauty that someone like Makara could see.

Within weeks money almost ceased circulating. There was little point in the material things of life any more. Only basic needs were filled and many people gathered in groups to talk and sing or just sit together. There were a few outbreaks of violence, but the rioters were quickly brought under control by unified peace-makers. The Followers' ranks swelled as never before. People's lives suddenly seemed more relevant to the broader universe than to their tiny doomed planet.

The small spacecraft headed successfully out of the solar system with its four occupants less than a year later. Nyra had withdrawn from everything in her life but Initiate work. Gradually she had resigned herself to the new mission, to carry the seed of flesh outward. But for her it was a mechanical task, even after the ship left orbit. What mattered was the spirit of Makara in her heart and mind. They *would* be together again. Were they not together even now?

The remaining Uranus-bound survivors rocketed off from little-known centers in less reliable craft, for the one place it proved impossible to quell riots was around the main launch center. Many of these ships plummeted back into the atmosphere in fiery death. Those who got beyond it were well underway when it became apparent to those remaining on earth that the nature of the sun's radiation was intensifying. Now even the Followers could not prevent mass hysteria and panic. It was worst in the big cities, which experienced a complete break-down in law and order. Since most people no longer saw any reason for working, the economic machinery came to a halt and the bare necessities of life became increasingly scarce. Social disintegration followed.

On the moon of Uranus the new struggle for survival was highly organized. Mineral resources had to be mined efficiently. Everything was on hand to build starships that could leave the solar system. They had a hundred years or so to do it. The human survival instinct had worked in a strangely co-operative way for the last several decades to get this equipment to Uranus, and now they could rely on the skill and luck of the selected survivors to get out of the solar system and find a new planet among the stars. They had to find a star that was not yet five billion years old. They chose one that was already four and a half billion years old, but at least it was stable, and it had planets. That was the one that the four-person starship was already headed for. The rest would reach Alpha Centauri thousands of years later. Undivided dedication was their only hope for survival. They all had to work for the survival of their children's children's children. As the sun became a red giant, their work proceeded unabated.

The best current estimate for the life expectancy of our sun is about five billion more years. That may seem like an unimaginably long way in the future—but what if our time estimates are wrong?

The sun, like most stars, will end its life as a planetary nebula. This relatively gentle death, however, is preceded by a red giant phase that will destroy all life on the planets, even swallow up entire planets.

The red giant phase begins when the hydrogen in the core of a star runs out and the star is forced to burn helium. The temperature is greater when helium burns, so the outward push of the heat overcomes the star's gravity. The star expands and its

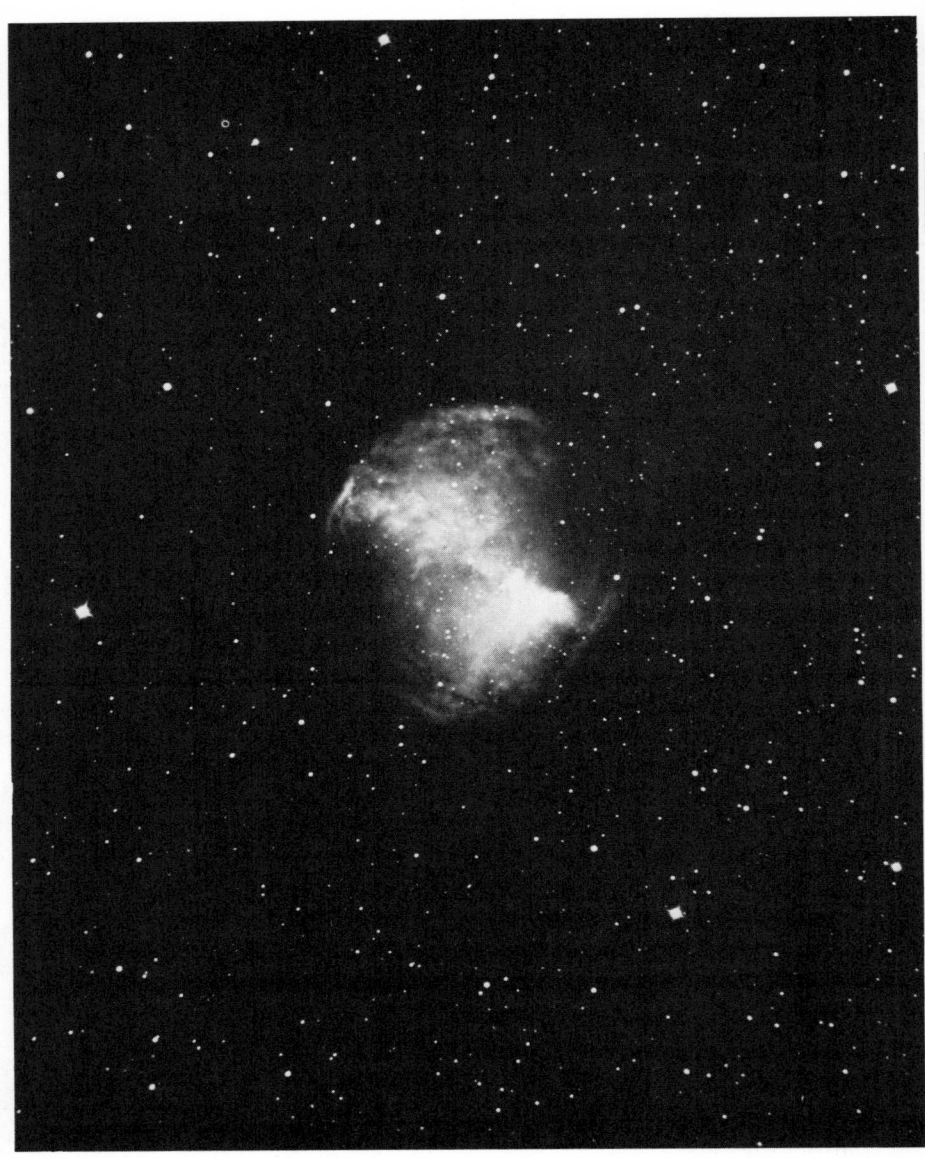

The planetary nebula known as the Dumbbell, in which the central star has thrown out a shell of matter in a final dying gasp. The sun, after it has become a red giant, will also pass through the planetary nebula phase, which should terminate any life still extant in the solar system. Hale Observatories photo.

surface cools. As it expands, it swallows up many of the planets orbiting it.

Any planets located beyond the outer limit of the ever-swelling star would be baked by the enormous increase in heat reaching them. The star's surface might be somewhat cooler than originally, but its surface would be much closer to the orbiting planets. The atmosphere of any planet not directly consumed by the red giant might be evaporated away by the intense heat. The surface temperature on earth is expected to reach 1500° C when the sun becomes a red giant.

All life in a red giant's planetary system would surely cease unless a civilization manages to fly to the outermost planets and set up new colonies there. If we were to do this when the sun starts changing into a red giant, our refuge would last only a few tens of thousands of years at best, because then the red giant would shed its outer shell of matter as it goes into its final death throes. This phase is known as a planetary nebula. Now the shell blows outward and sweeps the atmospheres off any remaining planets. If life hasn't already left for safer star systems, it will end. All living things in the solar system will be totally wiped out.

The old star gently sheds its outer layers (gently in comparison to a supernova explosion) and as the shells are cast off, the star reveals its hot interior.

We can see planetary nebulas all through our galaxy. The central stars are usually clearly visible, and their radiations keep the expanding shells luminous for tens of thousands of years. A typical planetary nebula appears as a ring of luminous material that is brightly colored when photographed through large telescopes. The shells move outward at about 10 to 20 kilometers per second. Ultimately, the central stars will cool down, fade, and die. In all likelihood, there were once planets in orbit about those stars. Many of them may have evolved intelligent life, but would it have survived the red giant phase? No, most certainly not. However, the ejected shell of a planetary nebula contains many of the elements built up inside the original star, and these can be used to form future stars and planets.

One thing is certain. Our scenario lies sometime in earth's future. Inevitably the sun will some day die, and all life in our solar system will die with it. But astronomers agree that this final catastrophe lies in the very distant future. Surely in the next five billion years we can learn how to colonize the stars.

HOW TO COPE WITH CATASTROPHE

CHAPTER THIRTEEN

Many chapters in this book have pointed up the dangers that life on a planet faces, and you may well ask how we have managed to survive so long. How did we make it to what we call the twentieth century without being destroyed? Have we just been lucky? If so, that would make us one of the few intelligent life forms in our galaxy. Or are there some crucial forces at work that ensure the survival of life once it gets a foothold on a planet?

There is a theory that might be used to explain how life can survive despite enormous odds. Perhaps it can account for our continued existence. Known as the Gaia hypothesis, after the ancient Greek earth goddess, this theory relates the atmosphere intimately to life. The originators of the theory, James Lovelock and Lynn Margulis, propose that the atmosphere appears as peculiar as it does because it is part of a living entity. We should regard the biosphere plus the atmosphere as a unity. Humans are then but a small element in this living organism that covers the surface of the earth. And the Gaia concept of a living atmosphere allows us to speculate how life has survived nearby supernova explosions, magnetic field reversals, and ice ages.

Scientists have long been accustomed to considering the biosphere (the system of living things on earth) as separate from the system of gases (atmosphere) that surrounds our planet, from the rocky body of the earth (lithosphere), and from the oceans, seas, and rivers (hydrosphere). While these latter three are needed to support life, they have rarely been regarded as part of life.

Our atmosphere consists of 80 percent nitrogen and 20 percent oxygen, with a little water vapor, carbon dioxide, and rare gases (like neon) thrown in for good measure. However, when we compare the earth's atmosphere with that of Venus and Mars, it is so strikingly different that we are forced to investigate why this is so. In the simplest view of the solar system, we might expect that we could infer the nature of the earth's atmosphere by comparing Venus and Mars and figuring out what a planet in between them should be like.

Both Venus and Mars are devoid of life, as far as we now know. We also know their atmospheric compositions very accurately. Using our knowledge of physics and chemistry, we can calculate what should have happened to the earth's atmosphere during its 4.7 billion year history if there had been no life present here either. Simple calculations lead to the prediction of an atmosphere containing a lot of water, a little carbon dioxide, and small quantities of nitrogen and oxygen, quite unlike what we now find. The oxygen would come from the breakup of carbon dioxide and water vapor by sunlight. The total atmosphere would only be about a third of what we now have (as far as pressure at sea level is concerned). Obviously, earth's atmosphere is not at all as we would expect from basic theoretical considerations, and we turn to the existence of life on earth to explain the difference.

It may seem obvious that our atmosphere is affected by life, but that doesn't make it part of a living thing. The Gaia hypothesis was only proposed after Lovelock and Margulis considered another particularly remarkable aspect of the past evolution of the earth.

There have been only small temperature changes on our planet, as inferred from the fossil and geological records of ice ages in the past. We also know that the earth has been covered by liquid oceans for at least three, possibly four bil-

Gaea. Oil on canvas. An enormous conglomeration of stars, a quasar or a giant galaxy, looms in the background as a symbolic tree of life surrounds and protects the earth among its roots, as a comment on the Gaia hypothesis.

lion years, but that at no time during the last four billion years have the oceans been completely frozen. Even during ice ages, with their temperature drops of 10° to 20°, the earth's seas and air stayed at a more or less constant temperature. This is particularly remarkable when we consider that the sun is believed to have changed in brightness and heat output since the earth and its oceans were formed. Estimates of past changes in solar brightness suggest that the sun's radiated energy has increased by 50 percent during the last 4.7 billion years.

Somehow life and the atmosphere (Gaia) must have acted together to keep the temperature of the earth constant. It also suggests that we, as a single species on this planet, despite the dire warnings we have been giving each other, are unlikely to be able to destroy this planet by our pollutions and general interference with the status quo. Gaia sees to that. While we might be quite capable of killing ourselves, and perhaps some other species as well, we are not likely to ever wipe out all terrestrial life. Interactions within Gaia, between the atmosphere and the biosphere, will always restore equilibrium. If the sun, magnetic field reversals, and supernovae haven't already killed off all life on earth, then we humans aren't likely to succeed either, simply because we are such a small part of the biosphere. We might cause a local irregularity, but that's all. The earth, and the universe, are quite capable of continuing without us.

A second remarkable aspect concerning the constancy of the temperature of the earth relates to the changes that have occurred in atmospheric composition over the last several billion years. When the earth was first formed, gases escaped from its interior, and these gases formed the first atmosphere: mostly ammonia, methane, and hydrogen. There was no oxygen around then. The first living things, between three and four billion years ago, needed no oxygen for their survival. They thrived on the ultraviolet energy from the sun. We know this from the study of fossils from those times. The non-oxygen atmosphere then changed to one in which oxygen became abundant, principally helped along by photosynthesis in plants. About two billion years ago, the atmosphere became something like the one we know today. Now, virtually all life depends on the oxygen in the atmosphere in some way. Countless species must have gone extinct during the transition time from the non-oxygenic atmosphere to our present oxygen-rich atmosphere. Yet despite the enormous change in chemical composition, the atmospheric temperature stayed constant.

It is a well-known physical fact that when an atmosphere changes its chemical composition, it also reacts differently to sunlight. It absorbs different amounts of light, heat, and ultraviolet depending on its chemical make-up. Therefore, even in the simplest picture we can conjure up, the temperature balance of earth's atmosphere should have changed substantially when the atmosphere changed its composition. Yet it didn't. The temperature stayed constant to within 10 degrees or so, despite the enormous chemical changes. What was responsible for that? The answer suggested is that a regulatory process, much more complex than we have been able to imagine up to now, must have been at work. This suggests the presence of a living entity on a large scale, a scale that encompasses the whole earth.

There are many factors concerned with life and the atmosphere that could have played a role in the temperature regulation. In their description of Gaia (in a paper on this subject that appeared in the scientific journal *Icarus* in 1974), Margulis and Lovelock list some of the factors that help control the temperature of earth's atmosphere.

They suggest that when looking at the past history of the earth, we imagine an engineer who has been given the task of keeping the earth's atmosphere at a constant temperature. In the face of changes in the sun's energy output, he might suggest that we change the amount of heat radiated back from the earth. White surfaces are known to be good reflectors, while black surfaces are good absorbers (or radiators). Other colors, such as green, red, or blue, have their own absorption and radiation characteristics. By controlling the color of the planet we would have a simple method (in principle) of maintaining good temperature control. Even if there were a lot of snow on the planet at some time, the amount of light reflected by the snow could be altered by small organisms living in the snowflakes. So life, in its various forms and colors, can significantly alter the temperature of the planet and hence can be used to regulate that temperature, provided it can respond in some subtle way to external changes in the heat or light reaching the earth. The known temperature constancy suggests that this may have happened.

Another factor that may have operated concerns the amount of heat absorbed from sunlight, in particular the amount of heat trapped in the atmosphere. This depends on the chemical composition of the atmosphere itself. We know that Venus has an incredible surface temperature, about 425°C. Its thick carbon dioxide atmosphere (100 times as dense as the earth's atmosphere) traps a great deal of heat beneath its cloud cover and acts like a planetary greenhouse. The earth, with a little carbon dioxide in its atmosphere, exhibits an active greenhouse effect that now keeps us some 20 degrees warmer than we would otherwise be. Without this greenhouse, our planet would be largely frozen. Ammonia is another gas that can produce a significant greenhouse effect. Presently, the generation of ammonia by animals balances its use by plants, and most of the ammonia is quickly removed from the atmosphere. But the net effect of the carbon dioxide is to keep our planet at its mild temperature.

In the event of an outside-induced change in the temperature of the earth's atmosphere, the production of carbon dioxide or ammonia by living things could act to change the greenhouse effect. It suggests an additional way in which life itself can act in conjunction with the atmosphere to keep things at a healthy temperature.

We might also consider other processes that can act to change the temperature of the earth. The amount of particulate material in our atmosphere can alter global temperatures considerably. At any time 50 percent of the earth is covered by clouds. Change that significantly, and you would change the temperature of the whole planet. Cloud cover depends on evaporation, which in turn depends on cloud cover, and so on forever. However, evaporation can also be controlled by making the surface of the water less likely to evaporate. This can be done by altering its surface structure. For example, it is possible for a scum layer to cover the surface of a lake, or perhaps even a sea, and hence reduce evaporation. This would change the local humidity, cloud layer, and other local features. In principle, even organisms living in water could alter temperature on a planetary scale.

Now consider the presence of particles in the air. These can abosrb heat; hence, depending on their presence or absence, and their sizes, they can change the amount of heat actually reaching the surface of the earth, thus changing the amount of evaporation. Examples of such particles even in prehistoric times would be those ejected from volcanoes and the smoke from forest fires. According to the Gaia hypothesis, these may also play a role in sustaining the overall temperature balance of the planet.

A delicate balance has obviously been maintained for so long (over four billion years) that all of these mechanisms may have acted for the survival of earth life. Gaia probably explains why we are here after many changes in the atmospheric composition, nearby supernovae, and changes in solar brightness.

During the history of the earth, our planet must have been within 30 light years of a nearby supernova maybe a dozen times, and on occasion in the past as many as 70 percent of all species on earth have gone extinct, and yet we are still here. Why? Possibly the answer is that we cannot consider life on earth independently of the planet, especially its atmosphere. An increase in cosmic ray intensity as a result of a nearby supernova would certainly produce an initial dramatic change in the ozone layer, but in the presence of Gaia this reduction in ozone might be countered by a reduction in the number of organisms naturally producing nitrous oxide, the substance that can destroy ozone. A balance in ozone production/destruction might be quickly reached again. This would prevent the total destruction of life so that while many species might be wiped out, there would always be some that made it through the crisis, with a little help from Gaia. According to the hypothesis, Gaia acts for survival as the whole system of biosphere, atmosphere, and probably hydrosphere respond on a planetary scale to changes in the environment.

This remarkable theory, which suggests that life on earth, together with the atmosphere, is in essence a single living entity, raises a final question. Where do we draw the line between life and nonlife? Is life just here on earth, or do we draw the line just outside the earth's atmosphere, or somewhere out in the universe? Previously, the line has been drawn at the exterior of the bodies or structures that one called living things. Now it appears that, taken as a whole, the entire planetary surface may be one living thing, including that very thin layer we call the atmosphere. In other words, just as we have organs to help us breathe, clean blood, and circulate oxygen, so the planet has organisms that help keep it alive as a whole despite outside influences. In a sense, we all help keep Gaia alive. The bacteria, the plants, and the animals are all parts of this large living entity. Remove one species, and in some subtle way the removal will only interact with other parts of the system to ensure the continued survival of the whole. If we start polluting the air with our vapors, fumes, and dust, there will be counterbalancing effects produced by Gaia that will change the life structure of some individual organisms, perhaps ourselves included. But life will go on, and the planet will survive. The boundary to Gaia is clearly above the atmosphere, but is it beyond that, out in space somewhere?

Perhaps life's boundary is well beyond earth. For example, the magnetic field of the earth may be integral to life. Its absence or presence affects life on a planetary scale. The earth is constantly moving through the solar wind, an extension of the atmosphere of the sun, and the sun is absolutely basic to life, as the fundamental source of life-giving energy, so can we really consider the sun any more separate from our life than our atmosphere? Where do we draw the line? At this point there is no obvious answer, but it is clear that life on this planet is part of a much larger scenario than we have heretofore dreamed. As we continue to study the universe about us, we find that we are related to it, and part of it, in ways that perhaps only the mystics have appreciated until now.

CONTACT!
CHAPTER FOURTEEN

On board the space station in earth orbit all was quiet as the crew waited for the shuttle to bring in the relief crew of astronauts and scientists. For the past four years the shuttle mission had been a resounding success. In particular, the large space telescopes, now in position, had extended human vision into the depths of space and allowed astronomers to study new cosmic vistas. Astronomical knowledge was being revolutionized, and the most dramatic event had been the direct photographing of planets orbiting a dozen nearby stars. Although they appeared only as specks of light, they had at least proven the existence of other planets.

It was while the LST was pointed at one of these stars that a faint trail was noticed on the electronically recorded photographs. A quick check ruled out a possible asteroid or comet. This new object's orbit was soon established and the scientists on board the shuttle realized that whatever was producing the trail was not in a simple orbit about the sun. It was headed very nearly straight toward the sun, but most incredible of all it was shining by reflected sunlight. It didn't take the astronomers long to realize that the object was not very different in size from their own shuttle craft and about as reflective. The solar system was apparently being visited from space by an object that must have had an alien origin!

The discovery wasn't very old before it was splashed all over the newspapers on earth. Plans were hastily made to use one of the shuttle craft to venture out into the solar system to try to intercept the alien craft. Scientists and engineers had about a year to make rendezvous. After that, the visitor would be lost to the grip of the sun. For once all bureaucracy was set aside as everyone cooperated on a grand scale and a modified shuttle, manned by three United States astronauts and one Soviet cosmonaut, blasted out of earth orbit on September 29, 1984.

Tuesday, December 17

Robert Craft took his eye away from the telescope eyepiece and rubbed it. Then he moved over to let Alexandr Sorin take a look.

"There's no question about it. That is a spacecraft of some sort. It looks about as alive as any of our orbiters or Pioneer craft of the seventies; in other words, it is inanimate." He spoke to his companions and into his microphone.

"We copy you, Robert." It was Houston Control, who had just reported that the Arecibo radar had confirmed the extremely high reflectivity, proving the object was made of metal. Earth was soon to find out what it was that had entered the solar system from the depths of space.

"This is Contact Shuttle. We will proceed with establishing a rendezvous at 876 hours as originally

Beaver Pond. Oil on canvas. Humans now look toward the stars and distant galaxies in the search for life elsewhere in the universe.

scheduled, unless there is any sign of action from the craft. We are convinced that it is a remote device of some sort. Are you agreed?"

"We are, Contact Shuttle."

"This thing must have left its planet hundreds of millions of years ago, at the very least." Robert Craft peered through the telescope again, then continued, "Of course, if its builders had any ideas similar to what we had when Pioneer 10 and 11 or the Voyagers were scheduled to leave the solar system, we must surely hope to find a message on it. We'll do our best."

"We'll be with you, Bob. Here's hoping."

"It is clear that that craft is too large to take on board. Our only hope is to have enough time to dismantle something and take it back with us. We'll only have a short time to hang around." He pushed himself away from the telescope and let Sorin look again. The alien craft was now looming up at them. They could already make out that the craft was shaped in a functional way, similar to other spacecraft that don't have to consider frictional drag on their journey through space. This one must have been meant as a deep space operation. It had a very large, disk-shaped antenna that dwarfed its main body.

The Contact Shuttle crew had about three hours to perform their EVAs and maneuver themselves close enough to the alien craft to detach whatever they could. Their equipment consisted of powerful laser guns, which could be operated only if they could attach them to the spacecraft, and many hand tools that would allow them to hack at the surface or even open something up, provided there was something to be opened up. They also carried a supply of space bags, which were containers that could hold hundreds of kilograms of material. These bags were equipped with small gas jets that would drive the salvaged material away from the alien ship. The space bags would be placed in a trajectory that would allow the shuttle to rendezvous with them later on. Beacons on these space bags would facilitate easier retrieval.

The mission was dangerous. They would have to rely on their spacesuits and their personal gas jets to get them into a rendezvous situation. In addition, they had a large transport control unit, linked to their shipboard computer, that would allow them to get close to the alien ship quickly and in a more controlled manner. They would have to return to the shuttle hanging onto this device when the mission was completed. One astronaut was to stay on board. Even if the other three lost their lives, he could salvage the space bags and return to the distant orbiting space station, now over a million miles away.

Robert Craft, Alexandr Sorin, and Mike Lancey left the airlock at precisely 7:25 A.M. Greenwich mean time, watched by two billion people on a TV hookup. They strapped themselves to the transporter, and as it slowly moved away, they made the necessary course changes to bring themselves into a trajectory that paralleled the alien craft. It was now so close that it could be seen with the unaided eye.

As they approached it, cameras monitored their progress. On earth, reporters, commentators, and self-styled experts on extraterrestrial intelligence spoke to the watching billions. This television eavesdropping had almost been thwarted by the military and security minded of both hemispheres who wanted to perform the rendezvous in perfect secrecy, just in case something dangerous was discovered. Their opinions had been overruled by the well-informed populations of the planet, who had no intention of being left out when contact with an alien spacecraft was first established.

Even as the two astronauts and the cosmonaut

moved toward the alien craft, the giant computers back at Houston were making a new calculation of the orbit of the alien ship. At the desk in his tiny office, Jeremy Toth scribbled furiously as he took data from the computer and looked up numbers in tables spread about his floor and desk. The alien craft appeared to be close to a highly elliptical course that could have kept it in orbit about the sun forever, if only it weren't coming so close this time.

Out in space, 10 million miles away, the three lone men tensed as the maneuver was completed and they were finally moving parallel to the alien craft. They could cover the last kilometer with their backpacks. Robert Craft was the first one to touch the alien vehicle. By now the presence of its rocket motors was obvious. The hull of the craft was pitted and appeared badly damaged in places, but there was little time to look it over more than casually. The three drifted about the craft, and with two TV cameras and one movie camera they filmed all the details before they started on their demolition mission. Astronaut Lancey turned on the lighting unit strapped to his back so that they could film the dark side of the slowly rotating craft.

When the pictures were relayed back to earth, a crew of Houston scientists watched the images on giant video screens. It was their job to make recommendations as to which pieces of the alien craft could most easily be removed, even while the survey photos were being taken. Their decisions turned out to be straightforward.

Several fragile-looking members were obviously used to attach scientific instruments to the craft. They were to be snapped off immediately and the space bags filled as quickly as possible. Then the three suited men were to explore closely and see what else they could find.

Back in Houston, missing all the excitement that he could easily have joined by walking down the corridor to the communal TV screen, Jeremy Toth puzzled over his numbers again. He reprogrammed his desk control, linked to the giant computer in the basement, and decided to experiment a little. It was a hunch that might pay off. His excitment grew as he began to suspect that he could demonstrate where the craft had come from.

The space bag was soon filled with bits and pieces of alien technology. The functions of this craft must have been similar to those of the typical NASA deep-space probes, some of which were even now leaving the solar system on their endless voyage into space. Perhaps, just perhaps, the aliens had also placed some message on board. Now that their prior mission was accomplished and the space bags had been set adrift, Robert Craft eagerly began to explore his alloted area of the space hulk.

"I'm clear Houston. Beginning to explore further. Do you have any suggestions?"

The usually imperceptible delay of several seconds seemed like minutes. "OK, Bob. You have the sunside section. Let us have good close-ups as you go in. Help yourself to whatever seems pertinent and take care." He moved along the craft very slowly, pulling himself along, clinging to appendages and cracks.

"There is considerable pitting on the surface, of a very small scale, Houston. Here is a good view of it." He aimed his camera and held it steady. The image of the tiny craterlike marks flashed up on the TV screens of earth.

Mission control gave the go-ahead to the other two to proceed with their free exploration, and all three were informed that they had only minutes left. Out of the corner of his eye Craft watched the space bags moving toward the shuttle. They were now controlled by signals from Dick White, who had stayed behind. He was responsible for the debris and would assure its return to earth.

Over the communications link he heard an exclamation of surprise from Alexandr. He had found what they were looking for. A small box, partly hidden by what appeared to be the mount for a radio telescope dish, was affixed to the hull of the craft. The box was clearly meant to be detached; and it was easily freed. As the box came loose, Robert Craft and Mike Lancey let out a whoop. Their exclamation was echoed over the planet as the TV picture flashed on several billion TV screens.

It was immediately obvious that the inscriptions on the box's surface were totally alien. But the aliens weren't that different from humans. They had hoped that their craft would be discovered and had sent a message to posterity. It would only be a matter of time before the scientists back in Houston interpreted it, and no doubt they would immediately initiate a plan to contact the originating planet by radio. If the alien spacecraft had been on its way for a billion years, then such an experiment would hardly succeed, because that distant civilization might long ago have died out. But at least the planet earth knew that it was no longer totally alone in the universe.

Hours later the shuttle began its long journey back to earth orbit. It would still take more than three months to get there, but their time in transit would be put to good use.

It was decided that the alien casket should be opened out in space. The find was too great to risk an accident on the way home. Even as they figured out a way to open it, everyone knew that there would be a complex message of some sort inside. They were not disappointed.

What was obviously a record of some sort was packaged together with a device meant for playing it. The four spacemen took a few hours to assemble the device, and the signals were played over the relay link to the Houston computers. Many copies were quickly spread around the planet so that deciphering could be done as efficiently as possible, with many minds tackling the job from different directions.

Jeremy Toth had been performing his experiments in the computer for weeks. He had played with orbits, locations of stars, and other data. Now he felt sure that his hunch was right, but the computer didn't provide the final proof he needed. As he tossed about in his sleep one night, the coding experts were making fast progress on the alien messages. The record was obviously loaded with information. Some of the data were in the form of sound; they included very complicated noise patterns that were probably pictures coded in a peculiar way. They finally found the key that was to help them inscribed on the lid of the five-sided box. The aliens counted with the base five and hadn't used a binary code. Quickly the pieces fell into place.

The first piece of information that was unraveled concerned picture data about their star system. This alien civilization inhabited the fifth planet of a nine-planet system. Their star had a spectrum closely similar to the sun, a not too surprising discovery since sunlike stars are common in the galaxy. As the deciphering continued, color images were suddenly available for viewing. The experts studing the photographs found images not only of the alien planet and its teeming life, but also of its neighbor planets. Two images, front and back side perhaps, of every planet were shown. Some 30 images of the aliens' own planet were shown. It was a very dry, dusty place, with the inhabitants living in isolated, pod-like structures, evidently built to protect them from their harsh surroundings. They had also set up colonies on several other planets, as well as their own moon, again using the same podlike structures. A reasonable picture of life on the alien world was quickly formed and word spread that NASA would soon make the details public.

A rumor leaked out that the location of the planet was known, at just about the same time that Jeremy Toth realized his mistake. When he had fed the coordinates of Barnard's star into the computer he had neglected to take into account the proper motion of Barnard's star through space. That motion was well measured, and when the corrected data were fed into the computer, the answer came quickly. Now he knew where the spacecraft had come from. He immediately placed a call to the administrator's office.

The press conference was held outside the main administrative building. The images retrieved from the alien spacecraft were displayed for all to see, and everyone formed his or her own opinion about the origin of the spacecraft. One of the images was of a planet that looked vaguely similar to earth. That planet, however, was clearly not the alien's home. Several of the other planets were cratered, and one appeared very red. Several of the larger planets showed banded structures in their atmosphere. The viewers had an uncomfortable feeling that they were looking at a planetary system similar to their own. But there were some subtle differences hard to pinpoint.

The home planet of the aliens was very desolate compared to the earth. The creatures themselves were so alien looking that it was hard to figure out which photos were of the intelligent species as opposed to other animal life on that alien landscape.

The intelligent creatures were bipeds, and they seemed covered in layers of what might have been skin, although it could have been their clothing. The existence of two sexes could be recognized.

The NASA administrator stepped forward and stated he had an important announcement. One of their experts in celestial mechanics, Jeremy Toth, had succeeded in solving the orbital elements of the craft and had been able to state with certainty where the alien ship had been on its journey through space. It was in orbit about Barnard's star and the sun.

A breathless hush, usually reserved for only the most solemn occasions, fell on the assembled reporters and staff of the NASA laboratories.

The administrator continued. "What we have discovered is of the profoundest significance to the human race. I would like you to look again at the photographs of the planets in their solar system. Here they are." A giant video image flashed up on the TV screen behind him. There was a pause as everyone watched the images on the screen dissolve from one to the other. They were being shown the views of the other planets. The dissolve sequence stopped at the familiar looking blue-colored planet. There appeared to be white clouds on it, but no landmasses that looked anything like earth. Two views were then shown. Clearly both sides of that planet were represented, and the careful viewer could see two landmasses on what was otherwise a very watery looking blue planet.

"This was not the home of the aliens." The administrator's voice intruded into the thoughts of billions on earth who were staring at the earthlike planet. "But we do believe that this planet is our home now." There was a stunned silence. He continued. "Their planet was apparently number five out from the sun, here." Several images flashed by. "And here is their base station on the moon of that blue planet, our moon." The next photo looked exactly like the earth's moon, except that the next image was a photo of a small colony of pod structures inside a crater.

"We have also dated the material of which the spacecraft was made," the administrator went on. "It is no less than half a billion years old." He paused, and then continued. "This spacecraft has been traveling through space for that period of time. However, our orbit calculations show that it has in fact been in an extremely elongated orbit that has taken it past its parent star, our sun, and then Barnard's star, once every hundred million years. Somehow, it has suffered a slight deflection from that trajectory, and now it is falling into its parent star. We are certain that the home of this spacecraft was on the fifth planet of this solar system. The photos you have seen were in fact of Mercury, Venus, Earth, Mars, and the outer planets. The fifth planet today is Jupiter but this picture sequence makes Jupiter the sixth planet. Pluto is missing. It appears that 500 million years ago there was a planet where we now know the asteroid belt to be."

An excited murmur rose from the crowd of reporters and a number of them quickly dashed for the phones, even though TV coverage of this historic event blanketed the entire planet.

"It appears that nearly half a billion years ago there was another planet in our solar system, one that was inhabited by creatures that did not exist in the watery environs of earth, but who found it worthwhile to keep a close watch on our planet from our moon. Primitive life was teeming on earth then, but out on this fifth planet, life was highly civilized. They had evolved to an extremely advanced state and sent this spacecraft and probably many others into space before their planet was destroyed. We have no idea how it was destroyed. Perhaps Pluto is a remnant of that planet too, thrown out by an explosion and then trapped in a solar orbit in the outskirts of the solar system. Perhaps their survivors finally made it down to earth. We will never know.

"They obviously planned for their craft to orbit Barnard's star and then return to earth, perhaps in a controlled search for life. The messages we found were probably meant for a Barnard's star civilization. Perhaps, for some reason, they were expecting to find life there. We will never know. What they didn't tell us is just why the spacecraft was in this peculiar orbit. Before I open the meeting to questions I wish to announce that NASA will be planning extensive missions to the asteroid belt, the moon, and Pluto to see what else we might learn about the fate of the earliest known civilization in our solar system. Above all, we plan a manned mission to that crater to see what they left behind. A full analysis and report on the data from the alien craft will be handed out immediately following our gathering. Are there any questions?"

In a small crater on the moon, a thin layer of meteoric dust covered the artificial dome, now waiting for its first visitors in 500 million years.

OUR FUTURE
ON EARTH
CHAPTER FIFTEEN

Sometimes we forget that we are only part of an enormous universe, perhaps an infinity of universes beyond our senses. Yet it is abundantly clear that we are here only because of processes that have occurred, not only on our planet, but also in the universe as a whole since the beginning of time. The shape of our future, as well, will be molded by forces that are probably way beyond our control. Can increased scientific knowledge give us any insight into the future of life on earth, on the largest time scale?

Perhaps we should seriously consider some of the scenarios for future cataclysms outlined in this book and explore whether we really have the potential to survive any of them. Of course we humans, with characteristic assurance, believe we can survive any catastrophe (even the energy crisis), but remember that the sun will die one day, remote as it may seem, even if it is five billion years in the future. We cannot even speculate on where humans will be by then. The death of the sun will be the end of earth unless we have thought of ways to control the evolution of a star (a not very likely possibility, even allowing five billion years for us to figure it out) or have found a way to emigrate to another star system. Human life on earth will be doomed.

In an earlier chapter, we outlined a scenario of life on earth in the event that a supernova should explode fairly near to us. A supernova explosion of that kind could happen any day. It might be half a billion years before it happens, or it could happen next week. Perhaps Gaia would protect us. A magnetic field reversal, indirectly facilitating the destruction of the ozone layer and exposing us to harmful effects from the sun, could occur any decade now. Who knows? Glaciers could blanket the planet within a few thousand years or even within a century. At some time we might drift into another dense, interstellar dust cloud that would precipitate another ice age. Who knows? Can we cope with any of these events?

If the system of life on earth does act for its own survival, atmosphere included, then such dramatic, externally induced changes are likely to be countered by changes in the distribution of living things on the planet so as to keep the whole system near a balance. Although some life, possibly human life, may become extinct, life as a whole is likely to survive. What we need to consider is whether some specific changes might not hasten the extinction of human beings, as opposed to other animal life.

For example, a very severe ice age that covered most of the developed world, North America, Europe, and the USSR would surely cause chaos in technological societies as we know them. There appears to be no way to shift billions of people nearer to the equator where they might survive in greater comfort. There would be no way we could prevent the destruction by advancing ice of most major cities on earth, unless we started to tamper with the climate.

And who would dare predict that we are entering a thousand years of geophysical tranquility? No one! We have had it pretty good for a few thousand years, but we know it can be harsh for life on our planet, especially on the cosmic scale. Consider that the earth is still seething inside, seething with molten lava and with forces that still push continents slowly about. Consider that in the past giant volcanoes burst open to rain ashes over half the United States to depths of many feet. Consider also that you won't find a geologist anywhere who will console you with a statement that all is now quiet inside the earth and that such volcanic eruption could not happen again. Geophysical violence happens every month. It will continue to do so, but at the same time we are so lulled by the day-to-day security of our lives that an earthquake here or a new volcano there doesn't really make an impression on us. Sometimes, when a few thousand are killed, it makes the news for a few days. Then we forget.

Of course, it is possible that humans won't survive very long into the future in any case. Besides obvious terminators such as global nuclear holocausts, there is also the possibility that, as a species, we have already become too intelligent to survive much longer. Here, although it may sound hardhearted, I refer to our ability to keep people alive who would otherwise die. It used to be that evolution encouraged the generation of the fitter species, those suited to their environment. But as we continue to keep ourselves alive in the face of what would otherwise be a natural death, and as we continue to produce more and more humans in an uncontrolled population explosion, we might already be tampering with our future in an irreversible way. The planet cannot, naturally, handle unlimited numbers of people. We are fighting our natural environment rather than living in harmony with it. We can imagine a situation in which keeping ourselves alive becomes totally regulated by ourselves, and this situation might be very unstable, for any slip would spell disaster.

It doesn't take too much imagination to jump a century or two ahead. Two hundred years from now we might live entirely off processed food, and the struggle to fight cancer caused by chemicals in our food and the environment might be just keeping pace with the birth rate, unless of course we have only cancer-resistant people alive then. Imagine then that there is some giant power failure all over the planet. All the food processing plants break down. Mass starvation follows because there aren't any more farms, and there's not enough time to go back to tilling the land and raising crops!

These are only two of the many ways we humans endanger ourselves. If we do exterminate ourselves as a result of our cleverness in taking over the reins of nature, the planet as a whole will still go on living and breathing. It will just be devoid of one of its more aberrant creatures, the ones who took it upon themselves to take over too completely.

Prognostications about our future life on earth, then, in view of all the cosmic dramas that can play themselves out, and in view of our technological oversophistication, are likely to be less than optimistic. This should not be too surprising. Life on earth has been struggling against a hostile environment for billions of years, and millions of species did not make it to the present time. The dinosaurs are only the most famous of the creatures that have gone extinct in the past. Our future will be no easier than the future for any other species.

In looking at the past and guessing about the future, we suffer from a limited view of our history as thinking beings. We only have 3000 years or so of even partially documented history. During that time we have battled among ourselves, but we have never been threatened in any substantial way by forces outside our immediate environment. On a cosmic scale, however, this 3000 years is only an instant in time, and we have just been lucky. It is only 11,000 years since humans emerged from the caves after surviving the latest ice age, and who knows whether we can as easily survive another 1000 years and then last through the next ice age.

Our future, on a cosmic scale, will continue to be a struggle to survive, but we do have tools to help us in our struggle. We have our power of rational thought and our mastery of technology. You might point out that the cave dwellers had fire to help them survive, and from a cosmic perspective their tools were no worse or better than ours. However, it's hard to believe that we're not much better off than they were. Surely we can face the future with good heart.

The course we set, however, depends to a large extent on whether we think we have friends in the galaxy. Perhaps

Tree of Life. Oil on canvas.

there are lots of other civilizations like ours, all of which have survived countless cosmic catastrophes. They may also be looking outward eagerly and exploring the universe. If they should happen to contact us, and many people think this could happen any time, then we could learn from them and use their knowledge and technology to ward off cosmic catastrophe. However, this seems a forlorn hope, based perhaps on a religious need for guidance from outside ourselves. Until we have absolute proof that advanced intelligent beings exist elsewhere in our galaxy, it is surely up to us to find ways to survive possible crises, such as the coming energy crisis, and it is up to us to treat our planet as if it is the only one we have.

There are several hundred billion stars in the Milky Way. A very large fraction, probably the majority, have planets. If there were just one planet in each planetary system that were at the right distance from its star to allow our type of life to evolve, then there could be 10 billion earthlike planets in our Milky Way alone. None of them is necessarily inhabited by intelligent beings. There are many factors that have to be just right for life as we know it to emerge. The planet needs to be at a moderate temperature, perhaps it needs a suitable moon and a magnetic field, and the planet needs to be tilted in a way that allows for mellow temperature extremes. Its parent star probably has to be four to five billion years old, at least, for life to have reached our stage of evolution. The planet should not have been too near a supernova explosion at any time, or its life would have been wiped out before it developed sufficiently. Such ifs and buts could continue forever, because we really have zero information that will help us figure out how many inhabited planets there are in the Milky Way. Most astronomers believe that while there are perhaps billions of planets around, there is no clear way to estimate how many of these have developed life. In the meantime, let's not waste time hoping that we will find miraculous messages from space that will solve our problems here on earth.

Whether or not you want to believe that we are the only

ones in our corner of the universe, the reality of the matter is still that we do not know there are others out there, and in the meantime we should surely behave as if we are the only ones. We should assume that we really have a vital role to play in our galaxy. This is a profound thought for us to accept. However, many of the consciousness raising movements and philosophies of today are also instilling this awareness in us. The future lies with us. It is not out there in space somewhere, however much you want to believe in UFOs or other manifestations of alleged extraterrestrial intelligence. The planet earth may just be the most advanced planet; it may harbor the most intelligent lifeforms around in the sense that others are too far away to be of use to us.

Accepting that we may be alone would mean that it might be we who will explore space; it might be we earthlings who ultimately spread to other stars and planets and perhaps colonize them, and it might be ourselves with whom we will someday communicate over interstellar distances. But are we doing anything to bring that day closer? Are we really living on earth as if we were the only inhabited planet? Certainly not!

Instead of performing such exciting adventures of the human spirit as the exploration of distant space and learning to communicate with other obvious intelligences on earth (dolphins and whales), we still fight, enslave, and bicker. Admittedly, there have been some changes since the Dark Ages, but we are still at each others' throats as if we had a whole planet to waste in this endless battle. We are only just beginning to be aware of our place in the universe. When we begin to comprehend the true vastness of it all, in perspective, we will be filled with awe, humility, and an overwhelming feeling of responsibility.

Of course, we might destroy a large segment of the human race by any one of several means, in which case sophisticated space exploration would be impossible. These potential disasters include nuclear annihilation, overpopulation, and ruinous changes in the environment. One must suppose that the basic survival instinct is strong enough to avoid total self-imposed destruction, or life would not have evolved to its present level. But this may be too optimistic. Certainly, life could easily be set back by a natural catastrophe like an ice age, which would lead to an enormous reduction in the numbers of people available for continued technological progress. Being aware of these possibilities may foster a new perspective on our vulnerability as a life form on earth.

If we realize the truth of that and let our imaginations wander, isn't it inevitable that as we move into the future with the destiny of intelligent life on earth (and perhaps in the galaxy) in our hands, we should develop the option to be mobile, to someday move to where we can survive? That means we must go outward. We will reach the outer planets in manned spacecraft, and we will fly to the stars—not soon, but maybe a thousand years or so from now, if it's not too late then.

In the first 70 years of this century, we increased our understanding of the universe by leaps and bounds. Now the end of this escalation in knowledge may be in sight. An obvious barrier now looms ahead. Humans are about to run out of cheap energy. Also, the scale of the equipment scientists now use is so enormous that the costs are reaching a limit. It is always possible that national priorities will change so that the equivalent of the military budget of the major nations is given to science (with the military using the present scientists' share), but that doesn't seem likely. But how else will we be able to build bigger telescopes and particle accelerators? We apparently need much bigger machines to make really great strides in our awareness of what is in the universe, or inside the atom, but can any of that be achieved without cheap energy?

Perhaps the time is upon us, say within the next 50 years (ignoring the energy crisis), when pure research about the physical universe will be curtailed not just because of lack of funds but because the next step will be recognized for what it is, just another step on an endless journey. Particle physicists are already aware that perhaps there is no such thing as an end to their search for the smallest, most basic particle. Astronomers are beginning to talk of infinite universes. So where do we stop, or slow down?

Does the future hold a completely new direction for us, a direction that is based entirely on the survival value of what we do? Does it include some form of planetary control, for instance, or does it lead to a more mellow approach to life (and science) with a de-emphasis on technology? Who knows? It might not be too long before it becomes obvious that we need to reconsider our priorities.

If we were to be threatened by some major catastrophe, we would no doubt struggle to survive as a species, but the interesting question is whether we could do that without upsetting the larger balance of life and resorting to personal control of the planet. Would it be possible for us to survive independently of the rest of the earth, its life and its atmosphere, until the crisis had passed? It seems that nature would deal with that same crisis in its own way, and when we emerged from our shelters, conditions might be so different that we still couldn't survive without artificial means. We can draw this conclusion quite logically from the Gaia hypothesis. In the event of any substantial change in our environment, the only way we could be sure to survive would be to take over control of the planet, but that means we could never again let up on our control efforts.

Already, climate control is becoming big business, but control of this kind reminds me of the tightrope walker who cannot change his mind halfway across the Niagara. We might be able to change global weather some day, but that would also involve us in a never-ending battle with nature. In a few tens of thousands of years the Office of Planetary Control and Protection might have created a bureaucratic monster that would be totally in charge of our environment. We would always run the risk of getting behind in our control efforts, possibly with dangerous consequences. There would be endless tampering, just a little more here or there, just to survive.

Perhaps there are planets entirely operated, atmosphere and all, by intelligent beings. Once the planet is taken over, where does that process stop? Are the boundaries on the home planet, or do they go on to include the parent star and then the whole galaxy?

An alternative scenario for the future of human life would be to increase our awareness of the harmony and interdependence of all. We might then reduce our interference with nature's processes and allow it to operate in its own way as it has for billions of years. The universe has operated without humans for about 15 billion years. Our planet has been here for some 5 billion years. Now we intelligent creatures come along and, having evolved out of the chaos, start looking out and becoming aware. The questions are: Can we learn to live in harmony with the universe, or will we interfere and try to control it? Once we have taken control, can there ever be any turning back? Turning back might lead to immediate extinction. I, for one, would like to see us approach the future with much less effort to control natural processes. The universe has done fine without our helping hand until now. Of course, in the case of life on earth, many species have already gone extinct because nature did take its course. And knowing that there are cosmic catastrophes in store for us, and knowing that human-created catastrophes are even now looming ahead, we have to stop and consider what it means to involve ourselves in the destiny of life on earth.

We should take a careful look at our place on earth and our place in the universe, and see ourselves in perspective. We are but one species on one planet orbiting one star in one galaxy. We should also realize that other intelligent creatures are about to be wiped out by humans. Whales are an endangered species, and dolphins are not far behind. If at some time in the future we decide to lend a helping hand to the processes of nature by taking control of something so basic as the climate (as an example), then we should be sure it is a helping hand we are lending and not a destructive slap that will destroy us all. The concept of living on spaceship earth is surely valid, and it should be taken much more seriously in the future. It is the only planet we have.

INDEX

Italic numerals indicate illustrations.

Absolute zero, defined, 20
Adsorption of negative charges, 60
Alabama, meteorite in, 121
Amino acids, in meteorites and on
    earth, 129
Ammonia, in atmosphere, 180, 182
  in gas clouds, 54
Anatolian Gateway, 147
Antimatter, antineutron, and
    antiproton, 148–149, 152
Arizona, meteorite in, 121
Asteroid, 134
Asteroids and asteroid belt, 119
Atmosphere of earth, changes in,
    180–181
  composition of, 178
  dust pollution in, 98
  in Gaia hypothesis, 177, 184
  particle collisions with, 68
  temperature control factors, 181–
    183
Atmosphere of Venus, 182
Aurorae, 68, 73, 101, 104
Auroral storms, 101
Australia, meteorite craters in, 122
Axis, earth's, changing, 92

Barringer Meteorite Crater, 121
Beaver Pond, 188
Big bang, 19, 59, 154
Binary stars, 57–58
Biosphere, defined, 178
  Gaia hypothesis and, 184
Black holes, 23, 143–154
  event horizon, 144–145
  gravity of, 143–145
  pair production in, 148–149
  primordial, 152, 154
  rotating, 149–150
  in Siberia, 125
  time and space in, 145–146
  tunneling, 152, 154
Bolide, 119, 121
Bow shock, 68

Calcium, 54
California, meteorites in, 121, 126
Cancer, constellation of, 22
Canyon Diablo, meteorite in, 121
Carbon, formation of, 58
Carbon dioxide, in earth's atmo-
    sphere, 178, 182
  in Venus's atmosphere, 182

Carbon monoxide, 54
Change of obliquity, 92
China, meteorite shower in, 124–125
Climate, control of, 100, 207–208
    cyclic patterns in, 91–92
    increased brightness affecting, 97
    sun and, 91–107
    sunspots and, 101–106
Cloud cover, 182–183
Clouds. *See* Dust clouds; Gas clouds; Storm clouds
Coalsack, *53*
Comets, 99, 119, 124–125, 128
Constellations, *17, 22,* 151
Corona of sun, *105*
Cosmic rays, 27–30
    bad effects of, 28–29
    beneficial effects of, 59–60, 62–63
    defined, 25
    earth's magnetic field and, 67
Crab nebula, 25–*26, 63*
Craters, 118–129
    lunar, *120*
    on Mercury, *127*
Cumulous clouds, 60
Cygnus, *10,* 151
Cygnus X-1, 151, 154

Density waves, spiral, 95–97
Devonian Landscape, 70
Dinosaurs, field reversals and, 71–72
Droughts, sunspots and, 103
Dumbbell planetary nebula, *174*
Dust clouds, *53, 54, 56, 85,* 106
    ice age theory, 94–100
    lanes, *93*

Earth, age of, 208
    atmospheric changes, 180–181
    formation of, 50
    future of humans on, 199–208
    Gaia hypothesis and, 177–185
    Little Ice Age on, 100–106
    magnetic field of, 65–73
    orbit of, ice ages and, 91–92
    temperature constancy, 178, 180–181
    temperature control factors, 181–183
Earthbound Angel, 111

Earthquakes, 126
Eclipse, binary stars and, 57
    solar, *105*
Ecology of forest system, 62
Eddies in earth's core, 65–66
Einstein's theories of relativity, 146
Electricity, 60, 62
Electrons, creation of, 19
    in eddies of earth's core, 65
    magnetic field and, 27
    in neutron formation, 23
    positron and, 148–149
    proton storms and, 71
Elements, heavy, 21, 54, 58–59, 63
Endangered species, 208
Energy, from black hole, 149–150
    from cosmic ray electrons, 27
    forms of, 59
    fusion generating, 19–21
    sources, 62–63
    from supernova, 10
    See also Radiation; *forms of energy by name*
Ergosphere, 149–150
Evaporation, of black hole, 152, 154
    cloud cover and, 182–183
Event horizon, gravity at, 144–145, 152
    pair production at, 149
Evolution, of life on earth, 29–30, 91
    of stars, 16–31, 50–64
    technology and, 201
Explorer satellite, 67
Explosions, flare star, 27
    meteorite, 118–129
    novae, 25, 27
    on sun, 27, 69
    supernovae, 23–25, 50–64
Extinction of species, 29, 69, 71–72, 183

Field reversals, magnetic, 65–73
Fireball, 119, 121, 124
Flare star explosions, 27
Forest fires, lightning and, 62
    smoke particles in atmosphere, 183
Fundingsland, Stephen Rinn, illustrations, *4, 39, 51, 61, 70, 77, 111, 134, 147, 160, 179, 188, 203*

Fusion process, 19–21

Gaea, 179
Gaia hypothesis, 177–185
  conclusion from, 207
Galactic cooling theory, 94–100
Galactic year, defined, 95
Galaxies, 17
  defined, 18
  growth of, 20
  ice ages and, 91–107
  interactions between, 95–96
  M 51, 93
  Magellanic Clouds, 96
  NGC 2903, 153
  spiral patterns in, 93–98
Gamma rays, production of, 148
  wavelength of, 59
Gas clouds, 54–58
Glaciations, 91–107
Gravity, of black hole, 143–145
  at event horizon, 149, 152
  in red giant phase, 173
  star implosion and, 23
  starbirth and, 21, 54–55, 57
Greenhouse effect, 182

Hawking, Stephen, 148
Heat, black hole and, 152
  as energy form, 59
  infrared radiation, 145
  in red giant phase, 173, 175
  starbirth and, 21, 54–55, 57
  See also Temperature
Helium, in carbon formation, 58
  in early universe, 16, 19–21
  in interstellar gas clouds, 54
  red giant phase and, 173
Henbury Cattle Station, Australia,
  meteorite craters near, 122
Hercules, constellation of, 17
Hydrogen, creation of, 19–20
  in early universe, 16
  in earth's atmosphere, 180
  helium formation and, 19–21
  in interstellar gas clouds, 54
  red giant phase and, 173
  spiral patterns and, 94
Hydrosphere, defined, 178
  Gaia hypothesis and, 184

Icarus (journal), 181
Ice ages, 91–107
  dust-cloud theory, 94–100
  earth-orbit theory, 91–92
  Little Ice Age, 100–106
  survival through, 202
Ice epoch, 97
Implosion of star, 23
Infrared radiation, 145
Ionization, defined, 27
  of nitrogen atoms, 28
  of supernova shell, 27
Iron, 21, 23, 119

Jupiter, debris near, 119
  solar wind and, 106
"Jupiter effect," 126

Kirin Province of China, meteorite
  shower in, 124–125
Kitt Peak telescope, 85

Lava, orientation of, 66
Leonid shower, 128
Life, emergence of, 24
  evolution of, on earth, 29–30, 91
  future of, on earth, 199–208
  supernovae and, 50–64
  survival of, 177–185
Light, black hole and, 23, 143–145
  from cosmic ray electrons, 27
  from neutron star, 25
  spectrum of, 25
  speed of, 18
  from supernova explosion, 23–24
  wavelength of, 59
Light year, defined, 18
Lithosphere, defined, 178
Little Ice Age, 100–106
Lovelock, James, 177–178, 181
Lunar soil samples, analysis of, 99–
  100

M-16 nebula, 56
M 51 galaxy, 93
McCrea, W.H., 97
Magellanic Clouds, 96
Magnetic field, defined, 27, 65
  cosmic rays and, 30
  of earth, 65–73
  interstellar, 27–28

Magnetic field *(cont.)*
  of pulsars, 63
  "Magnetic tail," earth's, 68
Margulis, Lynn, 177–178, 181
Maria, 118
Mariner 10, *127*
Mars, climate alteration, 100
  debris near, 119
  craters of, 118
  solar wind and, 106
Maunder Minimum, 103–104
Mercury, craters of, 118, *127*
Meteor, 119, 121
  explosive equivalent of, 121
Meteor shower, 122, 124–125, 128
Meteoric dust, 124
Meteorite, composition analysis,
    128–129
  craters, 118–129
  defined, 119
  effects of, on earth, 126
  explosive equivalent of, 126
Methane, in atmosphere, 180
Micrometeoroids, 99–100
Milankovitch, 92
Milky Way, earthlike planets in, 204
  radio signals from, 28
  rotation of, 95
  spiral patterns in, 94–96
Molecules, building, 62, 129
  organic, in meteorites, 129
Moon, earth's, craters of, 118–*120*
  soil sample analysis, 99–100
Mutations, 29, 73

Nebula(s), Crab, 25–*26*
  energy in, *10*
  M 16, *56*
  Orion, *85*
  planetary, 24, 50, 173–175
  solar, 57–58
Neutron star, cosmic rays produced
    by, 27, 62–63
  of Crab nebula, 25
  defined, 23
Neutrons, antineutrons and, 148–
    149
  formation of, 19, 23
  in helium formation, 19
NGC 2682 star cluster, *22*
NGC 2903 galaxy, *153*

Nitric oxide (NO), 28–29
Nitrogen, in earth's atmosphere,
    178
  nitrogen oxides and, 28–29, 71
Nitrous oxide (NO₂), 28–29, 183
North Pole, location of, 92
Novae, 25, 27
  defined, 58

Obliquity, change of, 92
Ocean(s), black hole emerging in,
    125
  meteorite crashing into, 126
  temperature of, 178, 180
Odessa, Texas, meteorite crater in,
    122
Orbital cooling theory, 91–92
Organic molecules in meteorites,
    129
Orion nebula, *85*
Oxygen, in earth's atmosphere, 178,
    180–181
  formation, 58
Ozone (O₃), defined, 28
Ozone layer, cosmic rays and
    28–30
  earth's magnetic field and, 68–69,
    71
  Gaia hypothesis and, 183
  proton storms affecting, 71
  sensitivity of, 72
  supernova explosion affecting, 16

Pair production, 148–149, 152
Particulates in atmosphere, 182–183
Perseid shower, 128
Photosynthesis, earth's atmosphere
    and, 180–181
Planetary nebula, 50, *174*
  defined, 24
  red giant phase, 173–175
Planets, color and temperature con-
    trol, 181–182
  formation of, 54, 58–59
  life on, 50–64
  requirements for life on, 204
  in star's red giant phase, 175
Polaris (pole star), 92
Positron, 148–149
Precession, 92
Primeval Landscape, 61

Primordial black holes, 152, 154
Prominences, solar, 101–*102*
Proton storms, 69, 71–72
Protons, antiprotons and, 148–149
   creation of, 19
   in helium formation, 19
   in neutron formation, 23
   solar, 69, 71
Pulsar, 25, 62–63

Quasar, *179*

Radiation, as antimatter annihilates
      matter, 148
   at big bang, 19
   black hole emitting, 152
   from cosmic ray electrons, 27
   forms of, 59
   infrared, 145
   van Allen belts, 67, 71
   See also Energy
Radio signals and waves, from cos-
      mic ray electrons, 27–28
   from neutron star, 25
   wavelength of, 59
Radioactive decay dating, 66
Radioactivity, 29
Radiolaria, 69, 71
Red giant phase, 173–176
Redshift, 145
Remnant Bay, 4
Rosetta Cove, 77
Rubidium, meteorite dating and,
      128–129

San Andreas fault, 126
Satellites, artificial, 67, 72
Scenic Overlook, 39
Sedimentary rocks, magnetization
      of, 66
Self-Extinction of Homo Sapiens, 160
*Seventeen Million Light Years from
      Home, 51*
Shock front, 27
"Shooting star," 119, 128
Siberia, black hole in, 125
   meteorite event in, 122–124
Singularity, 146
Skylab, *105*
Sodium, 54
Solar flares, 27, 69, 72, 101

Solar nebula, 57–58
Solar system, birth of, 118, 154
   See also Sun; *planets by name*
Solar wind, 52, 100
   earth's magnetic field and, 65, 67-
      -68, 98
   ice ages and, 104, 106
Spacecraft, in future, 206
   photographs taken by, *105,* 118,
      120, 127
   See also Satellites, artificial
Spectroscopic binary, 57
Spectrum, indications by, 25, 57
Spiral galaxy, *153*
Spiral arms, patterns, and waves,
      93–98
Splash zone, *120*
Star(s), birth of, 20–21, 54–57, 96,
      154
   evolution of, 16–31, 50–64
   exploded, *10. See also* Supernovae
   implosion of, 23
   massive, collapse of, 150–151
   red giant phase, 173–176
   Trapezium, *85*
   youngest, 96
Star cluster, *22*
Static electricity, 60
Stellar winds, 50, 52
Stone, 119
Storm clouds, 60
Sun, *102, 105*
   brightness changes, 180
   climate and, 91–107
   death of, 24, 173–176
   explosions on, 27, 69
   formation of, *22,* 50, 57
   human life affected by, 100
   life expectancy of, 173
   Little Ice Age and, 100–106
   ozone layer affected by, 71–73
   ultraviolet radiation from, 28–29
   See also Solar wind; Sunspot;
      other solar phenomena by name
Sunspot, cycle, 101–106
   maximum, 72
   minimum, 103
Supernova(e), blast, 23–24, 55
   frequency of, 24–27
   frequency of earth's nearness to,
      183

Supernova(e) *(cont.)*
  life and, 4, 50–64
  remains of, *10*
  starbirth to stardeath as, 16–31
Supersonic transports (SSTs), ozone
    layer affected by, 71
Survival of life, 177–185

Temperature, atmospheric changes
    and, 180–181
  at big bang, 19
  of black hole, 152
  control factors for earth's atmo-
    sphere, 181–183
  drop during ice ages, 103
  earth's, constancy of, 178, 180–
    181
  earth's, during sun's red giant
    phase, 175
  fusion and, 19–20
  planetary changes, 92
  of a star cloud, 21
  of supernova blast, 23
  of Venus, 182
Texas, meteorite crater in, 122
Thorium, meteorite dating and,
    128–129
Tidal waves, 126
Time in black hole, 145–146
Transients, *105*
Trapezium stars, *85*
Tree of Life, 203
Tunguska valley event, 122–126
Tunneling, 152, 154

Ultraviolet (UV) radiation, 28–29,
    59, 180
  effects of, 71–73
Universe, age of, 208
  creation of, 18–20

Van Allen radiation belts, 67, 71
Venus, atmosphere and tempera-
    ture of, 182
Virtual pair, 149
Virus mutations, 29, 73
Visual binaries, 58
Vladivostok, fireball near, 124
Volcanoes, craters, 118
  lava orientation, 66
  particles in atmosphere, 183
  in United States, 200

Water, in interstellar gas clouds, 54
  vapor in earth's atmosphere, 178
Wavelengths of energy forms, 59
Waves, density, 95–97
  energy, 59
  See also waves by name
"White hole," 146
Winds, *See* Solar wind; Stellar
    winds
"Wormhole," 146

X-rays, black holes and, 151
  from neutron star, 25
  wavelength of, 59

Zero, absolute, defined, 20
Zero magnetic field, 65–73